# Demystifying Marketing

## Other volumes in this series:

# Demystifying Marketing

A guide to the fundamentals for engineers

## Patrick Forsyth

The Institution of Engineering and Technology

Published by The Institution of Engineering and Technology, London, United Kingdom

© 2007 The Institution of Engineering and Technology

First published 2007

The Institution of Engineering and Technology
Michael Faraday House
Six Hills Way, Stevenage
Herts, SG1 2AY, United Kingdom

www.theiet.org

**British Library Cataloguing in Publication Data**
Forsyth, Patrick
  Demystifying marketing: a guide to fundamentals
  1. Engineering services marketing
  I. Title
  620'.00688

**ISBN 978-0-86341-806-8**

Typeset in India by Newgen Imaging Systems (P) Ltd, Chennai
Printed in the UK by Athenaeum Press Ltd, Gateshead, Tyne & Wear

# Contents

# Preface

They say that if you build a better mousetrap than your neighbour, people are going to come running. They are like hell! It's marketing that makes the difference.

Ed Johnson

## Marketing matters

Marketing is an important and specialist activity, one that is key to success for commercial and noncommercial organizations alike. It is also one of the most misunderstood areas of business, particularly in the sense that the full extent of it is often underestimated and misunderstood to some degree by people in other functions and activities.

Yet it has been said that marketing is too important to leave to marketing people. Certainly its effects range far and wide and many people are involved, influence it – albeit obliquely – or are affected by it.

This book is intended to explain and demystify marketing, and to show how it works and how it affects people, specifically those who come under the general umbrella of the title 'engineer'. This is used, in the context of the Institution of Engineering and Technology, to encompass a wide range of people. Such include engineers in engineering companies, engineers contributing to companies aside from that definition and those working on a consultative basis and selling their expertise and time on a fee basis. This whole group runs a technical span from structural to chemical engineering, with much more along the way.

The book is not a textbook, nor is it primarily a manual for the practitioner, though those new to the marketing function or some part of it should find it useful to put everything in context, and it certainly takes a practical view of its subject. Its main aim is to explain marketing to nonmarketing people so that marketing and engineering activities will work better together because the people on either side of what can be a divide understand each other better. In explaining something of marketing (and its jargon), the book uses a variety of examples to help illustrate its content; some of these relate to engineering, others are chosen simply as common examples with which everyone is likely to have some familiarity in everyday life. Language too is kept general. For instance, the word *customers* is used, though a service organization such as a professional firm of consulting engineers would probably talk about *clients*. Although marketing is an area of some complexity, and certainly one with a great deal involved in it, the focus here is on those aspects of the process that are of most relevance for nonmarketing people in organizations where marketing is a key activity.

One idea aired here is that of what is called a *marketing culture.* That means that organizations in which everyone understands marketing, and if necessary their role in it, tend to relate better to their marketplace than ones in which this is not the case and are more successful as a result. Marketing people and marketing activity intend to create success, and thus often profitability; they will do so more surely if every aspect of an organization – and everyone in it – is aware of their importance and assists, where appropriate, with making them successful in an increasingly competitive world.

We start with an overview.

Patrick Forsyth
*Touchstone Training & Consultancy*
*28 Saltcote Maltings*
*Maldon*
*Essex CM9 4QP*

# Acknowledgements

First, it should be said that I am in a position to write a book such as this only because of the time I have spent in marketing consultancy and training. Over the years many clients and colleagues too numerous to mention have assisted me to forge a career in marketing and accumulate knowledge and experience of it – as well as some expertise in it. Thanks are due to all who have assisted me, wittingly or not, along the way.

As always I am indebted to those with whom I work on the publishing side of book production when writing. In this case thanks are due to Lisa Reading and her colleagues at the Institution of Engineering and Technology. Also to Beverly LaFerla, who, until recently, edited the Institution's publication *Engineering Management*, and who suggested that I might be able to produce something longer than the articles I have contributed to her magazine.

Patrick Forsyth

*Part I*

# An overview of marketing

In this first part, marketing, which can be a confusing area, is defined and demystified. The intention is to set out the whole of what marketing encompasses in the round, show how each of the many elements relates to the others and put everything in context. The overall business process, of which marketing consists, is described.

The goal of the first two chapters is a broad one: to show and explain the complexity of marketing, so that readers can understand the role of the various elements and see how they work together. Readers will also understand that both the overall premise of marketing, and indeed its component parts, are all essentially common sense. The broad view readers will have at the end of this section will ensure that, as different, and disparate, elements of marketing are subsequently reviewed individually, they will be able to appreciate them in context of the entirety of marketing and all that it embraces.

A subsidiary objective is to ensure that readers understand how important marketing is to business success, and also how widely it influences, and is influenced by, people and processes around the organization.

# *Chapter 1*

# Marketing in context

This first chapter sets the scene and begins to put the complexities of marketing in context.

## 1.1  Marketing misunderstood

Frequently over the years, when I am asked what I do then reply that I am in marketing, wrong assumptions are made. Other people do not have this problem. If someone says that they work as a dentist, for example, people understand; they may not like the thought of it, but they understand. In a company, too, the functions of many people are clear: the production manager keeps the production line moving, quality control speaks for itself, the accountant keeps the score, and the engineer oils the wheels – and, yes, much more.

But marketing? It is thought to be, well, sort of advertising – or selling. Often, people's first thoughts are followed by not only an inaccurate image but also a negative one – market traders, Brand X, supermarkets and too much 'junk' mail. Why is this? Marketing people believe marketing is important; if it is, then other people should know about it. Perhaps any confusion is the marketing people's fault. They should explain.

Does this matter? Yes, I believe it does, largely because so many people in an organization are, wittingly or not, involved in, or affected by, marketing. Many have an influence on its effectiveness. And this is true from top to bottom of the organization, in a whole range of departments and functions. So, it is important to any organization that marketing not be isolated. It helps if it is widely understood; indeed, in competitive times it *must* be understood if all are to play their part and marketing is going to be effective in 'bringing in the business'.

### 1.1.1   An overview of the nature and breadth of marketing and its role

Though marketing is to a degree a matter of common sense, it also involves complexity. For a start, the word *marketing* is used in (at least) five different ways. Also, the topics of all the sections of this book, which in a sense dissect marketing and review aspects of it individually and – to some extent – in isolation, highlight and explain that complexity.

Without a clear overview, what is described in the following pages is in danger of selling the various individual aspects of marketing short. It is this overview – setting out the broad picture – that this introductory chapter briefly describes. Its intention is to allow the reader subsequently to obtain more from their reading of the rest of the book once this overview is appreciated.

### 1.1.2   Scope and definition

Because of the confusion that sometimes surrounds marketing, we will start with a word about what marketing is not. It is not a euphemism for advertising, nor a smarter word for selling. Our first objective here is to demystify the word before looking at its relevance and application. Not only is marketing an area in which there is considerable jargon, it is itself a word that can confuse, because there is not just one straightforward definition of the word at all. The word is used in several different ways, and all are broader in scope and complexity than the euphemisms quoted above.

For any business, marketing first describes five things:

1.  It describes a **concept**, the belief that the customer is of prime importance in business, that success comes from customer orientation, seeing every aspect of the business through the eyes of the customer, anticipating their needs and supplying what they want in the way in which they want it; not simply trying to sell whatever we happen to produce.

    This is surely no more than common sense (though manifestly not something every business embraces in its entirety) and is something that certainly always has relevance. In different businesses, of course, the 'customer' encompasses a number of different people. Goods or services may be sold direct to the public, or to them through others (wholesalers, retailers, etc.); other sorts of marketing involve other types of customer, for example, business-to-business marketing, where, as the words suggest, one organization is selling to another.

2.  Marketing describes a **function** of business; to define it formally, 'the management function that is responsible for identifying, anticipating and satisfying customer requirements profitably'. In other words, it is the process that implements the concept. Such must clearly be directed from a senior level and take a broad view of the business. More simply put, someone must wear the marketing 'hat'. In smaller companies, this may not be someone labelled 'marketing manager' or whatever: the responsibilities may be with such people as a general manager or sales manager, promotion manager, or they may be – often are – spread around among a number of people. Whoever is involved and however it is arranged, the final responsibility must be clear and sufficient time must be found to carry it out.

3. Marketing is an umbrella term for a **range of techniques**; not just selling and advertising but all those techniques concerned in implementing marketing in all its aspects: market research, product development, pricing and all the 'presentational' and promotional techniques including selling, merchandising, direct mail, public relations, sales promotions, advertising and so on.

4. Marketing is an ongoing **process**, one that acts to 'bring in the business' by utilizing and deploying the various techniques on a continuous basis, and doing so appropriately and creatively to make success more certain. Marketing is not a 'profit panacea'. It cannot guarantee success, nor can it be applied 'by rote' – the skill of those in marketing lies in precisely *how* they act in an area that is rightly sometimes referred to as being as much an art as a science.

5. Now all this may well start to make things clearer, but it is still not the complete picture. Lastly, marketing has to operate as a **system** and involves variable factors that operate both inside and outside the marketing organization. Many are restrictions. After all, an organization cannot do just as it wants, ignoring the outside world – all sorts of factors might conspire to hinder their intentions. Such might range from competitive activity to government action.

The marketing system links the market (customers and potential customers) with the company, and attempts to reconcile the conflict between the two. A moment's thought will show that the objectives of company and customer are not the same. For example, the organization may want to sell its products or services for a high profit, whereas the customer wants the best value for money.

The details regarding the **marketing system**, which position the marketing process within a broader context and link the organization to the world outside, are picked up in Chapter 2.

At this point consider: What do you buy? Can you think of things (or shops and organizations that you buy from) that are less than perfect? Some may satisfy completely, but maybe service, quality or reliability could be better. Even though the concept of marketing seems no more than common sense, there is plenty of room for companies to compete by matching customer needs better than others.

So, what does marketing do to achieve its aims and lead an organization through the potential minefield of external factors that may influence it? A little more about the continuous implementation of the marketing process will fit the range of techniques into the picture. This implementation must, if it is to be successful, be executed in a way that keeps a close eye on external factors. This cycle of activity is shown in Figure 1.1 and starts, unsurprisingly, with the customer. As the process goes on, we can see how some of the classic marketing activities feature and how they relate to the concept in carrying out their specific role.

Let us consider in outline what each of these implies in turn.

• First, **market research** attempts to help identify, indeed anticipate, consumer needs: what people want, how they want it supplied, and whether they will want it differently in the future. Since research can analyse the past and review current attitudes, but not predict the future, it must concentrate on trends, and needs careful interpretation. Even so, it can have an important role in reducing risk and

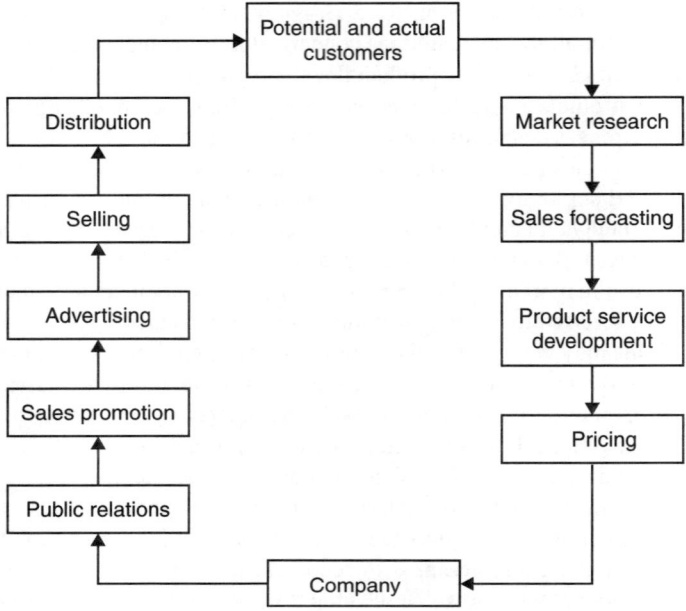

*Figure 1.1    The marketing process*

assisting innovation, and can be utilized throughout the marketing process, not just as a preliminary.

- Next, **forecasting** must be used to try to ascertain which quantity of a particular product/service may be purchased in future. Identifying a clear need is of little use commercially if only a handful of people want it. This is typical of areas where marketing does not offer exactitude. Forecasting is not easy – a point well made, albeit in another context, by the physicist Neils Bohr, who said, 'Prediction is never easy. Especially of the future' – and is never 100 per cent accurate; but the best estimate possible needs to be made to aid planning and reduce risk.

- **Product and/or service development** is, for most businesses, a continuous process. Sometimes the process is more evolution than revolution, since a product gradually changes; sometimes it is more cosmetic than real (a new, improved floor cleaner with ingredient X); sometimes it is so rapid that consumers get upset by the pace of obsolescence – as with computers, where it is said that you know that if you have the latest model it must be obsolete. No company can afford to stand still, and innovation in marketing, rather than the slavish application of the status quo, must be the order of the day, even if, for aspects of engineering, this can be a long-term process.

- **Price**, along with all aspects of pricing policy, is normally a marketing variable. Price says a great deal about quality and must be set carefully. This is not only to ensure that financial objectives are met, but also to create the appropriate image and feeling of value for money in the marketplace. Price is an inherent part of the product/service and must be used as an element of the marketing process.

With these factors in place, next we turn to external communications.

- Any company must **promote** itself; that is, it must communicate, clearly and persuasively to tell people what is available and encourage them to buy. A variety of techniques – advertising, direct mail, sales promotion and so forth – can then be used, together or separately. Of all the promotional tactics, selling is the only personal one, involving one-to-one communication, and it often forms a final, important, link in the chain of different methods that link a company to its market. The impact of visibility is clear if you look at the history of many products as they have grown from small beginnings to be market leaders in their fields, a process most often driven by the promotional investment made in them.
- Lastly, a further important part of marketing, not yet mentioned, is **distribution**, the process that allows products and services to be delivered within the marketplace. Marketing sometimes involves a direct relationship: you see an advertisement in the newspaper and reply direct to the company, who sends you the product. More often, there is a chain of intermediaries, as, classically, consumer products go from a manufacturer to a retailer before being bought by a customer. Such situations are duplicated in many industries, so, similarly, the new brake light bulb you buy for your car may go from the manufacturer to wholesaler to garage to you; other chains may be longer, and there is often both complexity and change with which to contend.

## 1.2   The company and its function

The precise way in which marketing fits within the way in which an organization is organized is also important. Every company has three basic functions – though in a well-directed company they do not operate in isolation from each other – and two major resources. The three basic functions are:

- production
- finance
- marketing

The two major resources are:

- capital
- labour

Each function has different tasks and different objectives, often operates on a different timescale, attracts different types of people and regards money in a different way. So, despite their all contributing towards the same company objectives, there is inevitably internal conflict between, say, marketing and production (and thus the amount of product it is thought should be produced and what may be sold), or production and finance.

*Table 1.1    How conflict arises between different company functions*

|  | **Finance** | **Production** | **Marketing** |
|---|---|---|---|
| Objective | To ensure that the return on capital employed will provide security, growth and yield | To optimize cost/output relationships | To maximize profitable sales in the market place |
| Time period of operation | Largely past – analysing results plus some forecasting | Largely present – keeping production going particularly in 3-shift working | Largely future – because of lead time in reacting to market place |
| Orientation | Largely inward – concerned with internal results of company | Largely inward – concerned with factory facilities for personnel | Largely outward – concerned with customers, distribution and competition |
| Attitudes to money | Largely debit and credit – once money spent, it is gone, money not spent is saved | Largely cost effective – hence value analysis, value analysis techniques and cost cutting | Largely 'return on investment' – money 'invested' in promotion to provide 'return' in sales and profits |
| Personality | Often introverted; lengthy training; makes decisions on financially quantifiable grounds | Usually qualified in quantitative discipline; makes decisions on input/output basis | Often extroverted; often educationally unqualified; has to make some decisions totally qualitatively |

Do note that organizations of course vary. *Production* implies factories and tangible products, but anything has to be produced. For example, software is not really in the same category as, say, a motor car, but it surely has to be produced. Even the team of people who audit your company's accounts represent the production side of the accountancy firm that does the work; a consulting engineer works in the same way. So, both products and services are involved here.

If you observe internal friction within your own company, therefore, relax – you are normal. Table 1.1 shows, in slightly caricatured form, how differences among people and functions affect the way things work.

Another piece of jargon, touched on earlier, that you will hear is the term *marketing culture*. This refers to the way in which marketing activity is supported and assisted by the attitudes and activities of people around the whole organization who are not actually a part of any marketing department, but whose work influences success in the marketplace. Maybe this includes you.

The figure suggests how main activity areas of the company need to understand each other and work together; the same is true of individuals at every level. As an example, think of people you have spoken to on the switchboards of companies you have contacted. In a few sentences, even a few words, they can influence you – for good or ill – in terms of how you think of their organizations and the likelihood of your doing business with them.

In many companies subcontracting is also involved and the activity can spread outside the company itself. Marketing must, therefore, work within the constraints imposed by the way the company functions operationally. Or must to some extent. Profits are, after all, generated only *externally*, and the organization as a whole must be organized in a way that allows marketing to be market-oriented. The market cares nothing for any internal inconvenience or confusion that may exist. It simply judges a company on its overall external image and the details that contribute to that. Thus, things done for internal reasons that do not work in the market may dilute overall marketing effectiveness and thus be damaging. Indeed, the 'sales' sometimes offer evidence of marketing people's overoptimism.

Since the 1970s, the marketing approach has predominated; indeed, increasing competition has meant it has become more important, thus become more sophisticated, and more concerned with the part of its activity designed to differentiate from competition.

## 1.3   The marketing mix

### 1.3.1   The three/four Ps (and more)

'The marketing mix' is a phrase, much quoted, that describes the *variables* that marketing people must work with in deciding on their chosen strategy. The phrase describes the various disparate elements of marketing, all of which are variable and must be organized in a way that creates an effective strategic approach: essentially it describes the product (or service), the price and how a message about what is offered is put over to potential consumers.

The marketing mix is often referred to as the 'three Ps', that is:

- product
- price
- presentation (or promotion)

These are all important and, of course, crop up throughout the book; the detail of them, however, is dealt with as follows. Product is the subject of Chapter 3, pricing of Chapter 4 and presentation is dealt with in the overall part of the book on marketing communications, a phrase that encompasses the whole range of promotional and sales techniques.

Sometimes a fourth P is added: place (this links to markets and distribution and is dealt with in detail in Chapter 6). Incidentally, it is worth bearing in mind the geographic span that the word *market* implies as you consider marketing. A small company may have a very local market (as a small retail business does), but it may literally involve the world – certainly everything that is said here applies as much to international marketing as anything else.

## 1.3.2    The extended marketing mix

Marketing people seem to have a love of acronyms, initialisms and other devices that encapsulate the many elements of their craft. Thus, the three Ps become the four Ps and, more recently, one has started to hear reference to the *seven* Ps. The extra Ps are all elements that have always been important, but they do slot into the overall concept of the mix and all are variables, needing decisions, and are factors that affect overall success, depending on how they are handled.

They are, in no particular order: *people, physical evidence* and *process*. Consider a few details about each.

**People:** The importance of people within the organization has already been touched on (the marketing culture). Here the concept is viewed more widely to encompass everyone at every stage of marketing activity, inside the company and outside: essentially customers, employees and suppliers. For instance, individual people – a waiter in a hotel or restaurant, for instance – have a major influence, and can almost be considered part of the product in a service. Similarly, chains of people all have to be satisfied and work well together if marketing is going to maximize its success. Here we might consider something manufactured: materials are drawn from various suppliers, designers and the like may be involved, subcontractors used (for anything from finishing processes to packaging) and then various people are involved down the chain of distribution (see Chapter 6). Marketing must work actively with all such people, and bear in mind:

- individuals' attitudes and how they influence performance;
- the quality of the motivational environment that keeps them productive;
- the skills they need and any training this may necessitate – again, clearly, the people aspect is a variable: the nature and degree of influence marketing brings to bear clearly can influence the level of market success enjoyed.

**Physical evidence:** This describes the tangible aspects of the delivery of a product to its customers. A good example of this is the merchandising and display that contribute to the convenience and visual impact of products on display in a retail outlet (the details of the techniques involved here are dealt with later). But suffice to say here that the intention must be to make purchase more likely.

**Process:** The kinds of process that this describes include the use of bar codes for product tracking and identification or the processing of a customer's credit card at the time of purchase. (A bar code is the block of black bars of different thicknesses printed on a product or its packaging. These can be read by the computers at cash points and help track and identify products: they put the price up on the screen and link to stock control to record another product sold and the number in stock reduced.)

Two things are important here: accuracy (customers will probably be upset if they are billed an incorrect amount) and time (customers expect all processes that go with purchase to be quick, convenient and not waste their time).

## 1.4   Summary

The whole process or cycle of marketing continues all the time, and already we see some of the different facets of marketing falling into place. As we will see it involves what is often referred to as being *as much art as science*. So:

- It is a creative process, one that has some scientific basis, but no absolute guarantee of success.
- The customer is always fickle and unpredictable; marketing may be an exciting function of business, but it carries a real element of risk.
- On the other hand, when it goes well it produces considerable satisfaction; this is a stage at which, with a product selling well, the marketing people tend to become convinced that the success is *all* down to marketing. In fact, as this introduction begins to show, a wider range of influences is at work. It is precisely because of this that it is often said that everyone in an organization is involved in marketing; and there is a good deal of truth in this.

So, marketing is much more than simply a department, or a body of techniques: it is central to the whole reason for an organization's being and to its relationship with its market and its customers. While, of course, many activities of a company are important, it is a truism that any kind of organization can create profits only out in the market. So, unless marketing activity, in the fullest sense of the term, creates a situation whereby customers buy in sufficient quantity, producing the right revenue and doing so at the right time, no business operation will be commercially viable. Marketing has to produce in customers a reason to buy, and make it a more powerful one than that which any competitor produces. Whatever the many elements involved, the key is to focus on customer needs and set out to satisfy them at a profit. Having avoided it early in this chapter, I deal in Panel 1.1 with the question of precise, overall definition of marketing.

### Panel 1.1   Definition of marketing

To summarize, and add a note of formality, here let me record that the Chartered Institute of Marketing has an official definition of marketing in its broadest sense that reads, 'Marketing is the management process responsible for identifying, anticipating and satisfying customer requirement profitably'.

The marketing guru Philip Kotler has defined it by saying,

> Marketing is the business function that identifies current unfulfilled needs and wants, defines and measures their magnitude, determines which target markets the organisation can best serve, and decides on appropriate products, services, and programmes

to serve these markets. Thus marketing serves as the link between a society's needs and its pattern of industrial response.

These certainly express something of the complexity involved. Marketing is more than just the 'marketing department', though the management guru Peter Drucker was content to say simply, 'Marketing is looking at the business through the customers' eyes' – and, indeed, everything stems from exactly that.

So, already, just with an introductory overview, it should be clear that there is more to marketing than may first meet the eye. This overview is designed to put the various elements in perspective before we move on and get into more detail on all the various aspects.

*Chapter 2*

# The marketing domain

If marketing happened in a vacuum it would be simple enough: find a product people want, make it and tell them about it. Of course, the reality is that there are many 'given' factors going on in the world and the market that affect any particular organization's intentions, and must be taken into account. Such factors can, in fact, help or hinder, but they must be recognized and worked with if marketing is to be successful.

## 2.1   The environment

The whole marketing system has to operate in an environment that may either restrict or assist it, but which certainly affects it. Such restrictions include:

- total demand
- availability of capital and labour
- competition (including international competition)
- legal requirements
- supply of raw materials
- channels of distribution, e.g. overseas agents and conditions
- technological impossibility

Any restrictions must be carefully considered, because of their effect on the business.

Consider these, briefly, in turn.

- **Total demand:** This is always finite. A greater number of people buy razor blades than rugby balls, and rugby balls last longer, too. Discovering the potential for any product is part of market identification and research, and marketing plans must always reflect this area of ultimate restriction
- **Availability of capital and labour:** Marketing clearly costs money, as does producing a product, but the details of corporate finance are beyond our brief here. Similarly labour, to an extent, though specialist staff may link very directly

to marketing (technical service support is just one example here that may involve large numbers of people) – good people help make marketing effective, a lack of them restricts it

- **Competition:** This is easy to recognize as a restriction. If competitor A sells more, then the market is reduced for competitor B. Few companies are monopolies (and, if they are, many governments will try to stop that situation continuing – or starting, in the case of mergers), but what is competition? Other companies making and selling the same product? Yes, but it is more than this. Take an example – a book, perhaps a novel or travel book rather than a business book such as this. With whom is the publisher competing? Other publishers of similar books are only the beginning. Competition is broad. The publisher is selling a product that fills leisure time, so is in competition with the theatre, records, movies, television, video and DVDs, magazines and newspapers. Developments in these areas affect its market. For example, how much has the advent of in-flight movies reduced the number of books passengers read and thus the level of book sales at airport bookshops?
- But competition is broader still. Book purchase comes from discretionary income; it is in fact not essential – though, come to think of it, I rate it as such! So competition comes also from other products entirely, the socks or pullover that need replacing perhaps. Expenditure on books is even reduced in a month when there is a particularly high telephone bill in the home of a regular book buyer.
- In addition, many books are given as presents, so items with a similar price that may also make attractive or appropriate gifts also feature as competition: pens, ties, costume jewellery and so on.

It may be a useful exercise to think through the extent of broad competition for another product, one with which you are familiar, or whatever your own organization may produce. The likelihood is that competition always proves broader than an initial, perhaps superficial, look suggests. Think about your own area of engineering and you will find the same thing.

- **Legal restriction:** We are all affected by legislation and legislation changes all the time. Marketers need to look ahead – what sales will be affected by a ban on fox hunting, for instance, and would such a change expand any other market?
- **Supply of raw materials:** Again this is a consideration beyond our brief in terms of the economics involved. One obvious example is the dependence of food products on the weather and harvest: the fruit available to make them limits the number of apple pies that can be sold.
- **Channels of distribution:** The lack of an overseas distributor in, say, Malaysia may hamper a firm's export, but it is easily recognizable and action can, potentially, be taken to correct the situation.
- **Technical impossibility:** In some fields of engineering the state of the technological art as it were may prevent, perhaps for the moment, something being done.

Restriction means just that. Some factors are at least bound up with the business and comparatively easy to work with. Other factors are truly external, and some act in the long rather than the short term. All can have direct impacts on markets and marketing opportunities – for good or ill.

Consider some further classic external influences. Watching the signs in all these areas may create opportunities as well as restriction, as the following examples illustrate:

- **Social:** Demographic trends (an ageing population in the UK) or lifestyle changes affect markets for products linked to, say, diets, holidays and health. So convenience foods present a greater opportunity when more families have both partners going out to work, and insurance can profit from providing travel policies for older people continuing to travel at an older age if they are fit enough to do so.
- **Political:** Regulations on safety affect both product design and price. Thus, regulations on safety affect car manufacture, and changes to controls on something like when a tyre has insufficient tread to be safe can immediately increase sales.
- **Technical:** A technological development such as the World Wide Web and email has created new product opportunities worldwide, certainly prompting the sale of more computers and the opening of Internet cafés (while email has no doubt reduced the market for post and sent the fax-machine market into decline). This too is an area that affects various sorts of engineering.
- **Economic:** Reduced taxes affect price and thus demand. Major economic developments such as those occurring as the EU reshapes itself can create new export markets or instigate other changes.

All the above are just examples. These are areas of considerable and ongoing change, and, as such, frequently suggest opportunities to those with an eye on the situation. They suggest difficulties, too, and – since forewarned is forearmed, as they say – this can help in taking action to minimize any adverse effects.

To demonstrate the possibilities here, take a few moments to focus on just one of the headings above. If you consider social changes, for instance, focus on yourself and your family and consider how your way of life has changed in, say, the last five years and how such change has affected what you buy. If things that you experience are common to many people, then the impact of them may be substantial and widespread. For instance, if you have a computer connected to the Internet, how has that changed your buying habits? Ditto other sorts of technological change in which you may be involved?

Many organizations separate the everyday aspect of managing activities concerned with marketing today's range of products from work on future projects. Unless sufficient time and resources – and therefore people – are spent in exploration and analysis (and in some industries, certainly aspects of the engineering world, extensive and technical research and development) as a discrete activity, the immediate pressures of securing current revenue can distract and lead to opportunities being missed or dangers encroaching unawares.

These kinds of influence both have general effects and also influence narrow sectors. For example, one current trend, referred to as GOY (getting older younger), has very specific influence on two sectors: many traditional toys are redundant as far as today's children are concerned at a younger age than in the past; conversely, other products are becoming appealing to younger age groups than in the past. Marketing must both respond to and deal with such factors and, many people believe, is also instrumental in encouraging them, even initiating them.

## 2.2    The product life cycle

The product life cycle charts the way in which a product performs over time from its inception and launch to such time as its decline may lead to its being discontinued.

### 2.2.1    The nature of the cycle

No product or service goes on being successful for ever. Some are here today and gone tomorrow, as with fashion products, a pop record or a newspaper or magazine. Research shows that, whether the life is short or long, over time the overall pattern is similar, taking the form of a bell-shaped curve, usually divided into five stages known as:

- introduction
- growth
- maturity
- decline
- phase-out

Let us consider these in turn.

- **Introduction** is a period of often slow growth as the product is introduced in the market. The profit curve shows profits as almost nonexistent at this stage because of the heavy expenses of product introduction.
- **Growth** is a period of more rapid market acceptance and substantial profit improvement.
- **Maturity** is a period of potential slowdown in sales growth because the product has achieved acceptance by most of the potential buyers. Profits peak in this period and may start to decline because of increased marketing outlays to sustain the product's position against competition. Much marketing activity is aimed at maintaining products at this peak and extending their life.
- **Decline** is the period when sales continue a strong downward drift and profits erode towards the zero point.
- **Phase-out** is the period during which, at some point, the product is withdrawn (or changed so radically that effectively its life cycle becomes a new one).

The designation of the beginning and end of each stage is somewhat arbitrary. Usually, the stages are based on where the rate of sales growth or decline tends to become pronounced. Not all products pass through the idealized bell-shaped product life cycle. Some products show a rapid growth from the very beginning, thus skipping the slow sales start implied by the introductory stage. Other products, instead of going through a rapid growth stage, go directly from introduction to maturity. Some products move from maturity to a second period of rapid growth.

### 2.2.2   Customers and the cycle

Support for the product life cycle concept lies in the way innovations are usually adopted into a marketplace. When a new product appears, steps must be taken by the company to stimulate awareness, interest, trial and purchase.

This can take time, and in the introductory stage only a few people (*innovators*) may buy it. If the product is good, larger numbers of buyers (*early adopters*) are drawn in. The entry of competitors into the market speeds up the adoption process by increasing the market's awareness and by exerting a downward pressure on prices. More buyers come in (*early majority*) as the product is legitimized. Eventually, the rate of growth decreases as the proportion of potential new buyers approaches zero. Sales become steady at the replacement purchase rate. Eventually, they decline as newer products appear and divert the interest of the buyers from the existing product.

Thus the product life cycle is closely related to normal developments that can be expected in the diffusion and adoption of any new product. Figure 2.1 summarizes all this graphically.

Think of a few well-known products. Some have been around a very long time – Bovril, Persil and Cadbury's Milk Tray chocolates. Some maintain themselves through modification, effectively starting the cycle again at its peak; others remain very much the same for long periods: for example, Black Magic chocolates were launched in the 1930s and did not change a single flavour for some fifty years. Others go into the doldrums, sometimes for years, but are revived, such as Brylcreem,

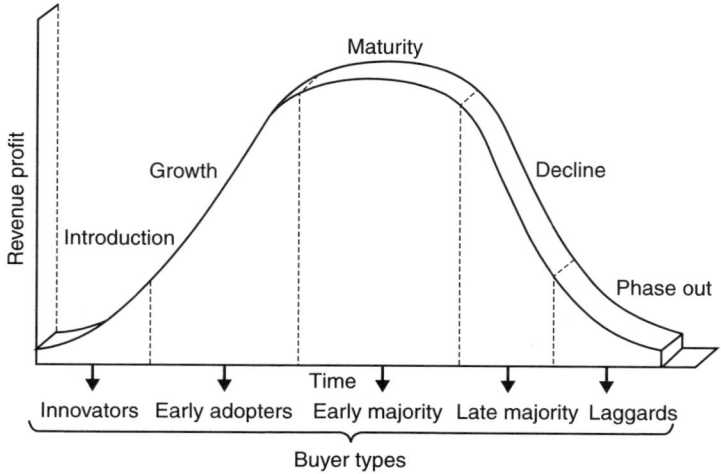

*Figure 2.1   Product life cycle*

Hovis and Lucozade; still others disappear without trace, for ever. All such occurrences are evidence of the product life cycle, and the more long-lived are usually a sign of successful marketing. You can doubtless think of more.

The product life cycle is something that marketers must respect; indeed the actions they must take are directed by it. Launching a new product is clearly a major task (and it is said that nine out of ten new products in the FMCG – *fast-moving consumer goods* – category fail in a short time). Beyond that, most marketing activity is directed to fit the stage of the life cycle at which a product may be. So, the marketing for a mature product is designed to maintain it at that peak and may contain elements designed to keep existing users buying regularly and also to attract new customers, each separate in the details of how it works.

If marketing deals accurately with this situation it is more likely to succeed.

## 2.3    Social and ethical considerations

### 2.3.1    The issues in question

Marketing is a force in the world. By its actions it can do good, helping consumers, providing choice and, in its social-marketing guise, literally aim to right wrongs and change things for the better. But it may also do harm. Consider two quick examples. If a product's packaging is, when discarded, causing harm to the environment, should the company concerned be prepared to spend more to avoid the problem? If a company's advertising uses images that offend some people, though help sell the product effectively, should they change this, even if it risks losing sales?

In the latter case many people might simply say that they should, but things are not so straightforwardly black and white. Taking the moral high ground poses many questions, for instance:

- Will customers notice such a change?
- What will customers think of the making of a change?
- What costs and effects on profitability will be involved?
- What will competition do?
- And, ultimately, what will be the effect on the business of one course of action rather than another?

Perhaps it is worth considering matters at three levels here.

1. **Seriously unethical behaviour:** This is still difficult to define, but much of it will be so bad that it attracts legislation. It would be wrong to sell untested medicines, for instance, and in most countries there are stringent measures in place to prevent it. Alongside this kind of thing, there are matters that are not actually illegal but are nonetheless subject to voluntary codes. While a poor medicine may kill people, a bad estate agent may only inconvenience (though they can cost their customers money if they behave unprofessionally), so here, as in many cases, there are voluntary codes administered by the industry. Some people might feel this is insufficient; indeed there are often moves to obtain legislation here. The lines

between such things are ultimately a matter of opinion. An unsafe product may not actually be illegal, but it is likely to attract regulation if it continues to be sold and do damage.

2.  **Borderline activity:** There is a middle area of marketing activity that stretches from practice that is, or has become, accepted to practice that is, if not over the edge, into the first category of seriously unethical behaviour. Some individual marketing practices span this range. One familiar example is what is called *confusion pricing* (more of which in Chapter 4). Essentially, this is what happens now with such things as mobile-telephone tariffs. The complexities have piled up and customers now find it almost impossible to make a true comparison among competitive offers. Which of us can put our hands on our hearts and say we know we have the best deal? This has become accepted as the norm, though it still annoys many and there is always the chance that one operator will appeal to those who are annoyed in this way with a different arrangement and create an edge on competition. The terms used to encourage people to switch credit cards are another example.

3.  **Welcome ethical behaviour:** I use this term to describe things of which customers, indeed the wider public, approve. Again, there is clearly a scale involved, but we all probably think it is right for companies to be environmentally friendly as much as possible, using recycled paper and promoting their cosmetics without unnecessary animal testing.

What is regarded as ethical or unethical may be difficult to nail down, but much is recognized by many people as being clearly on one side of the line or the other. Part of the motivation here for ethical and caring behaviour is personal. It would be nice to think most of those in charge – and in marketing – *want* to run their business that way simply because that is what gives them satisfaction. But the profit motive is strong and the amount of legislation and comment suggests that there are plenty of exceptions. Even so, some businesses are motivated in this way. A recent example is OneWorldHealth, the first nonprofit pharmaceutical company, an American enterprise aiming to provide low-cost medicines to developing nations. Perhaps there is a trend here that will grow.

### 2.3.2   *The effects of ethical/unethical marketing*

In practical terms, it is true to say that the way an organization conducts itself has a direct effect on consumer views of it; this is true also of the media. If things are regarded as sufficiently bad, the ultimate effect is a boycott. This has happened many times, and though it may have only gradual effects – such as a boycott on tinned tuna because of the way in which certain fishing methods killed dolphins – sometimes a long-term campaign is set up and can result in radical and permanent change. Started in the USA, the group United Students Against Sweatshops is one that is now a global force. Any dissatisfaction can lead some potential customers to vote with their feet and sales to be reduced.

Since we are looking at ethical problems directly stemming from marketing, it may be useful to flag some examples. The key here is less whether such things are 'unethical' in some measurable sense (some things are wrong, however you look at them), but whether a sufficient number of consumers feel they are bad enough for their subsequent actions – predominantly not buying – to affect sales. These go across the scale. Let us look at some examples.

- **Bribery:** Normally, this would be frowned upon, and there are often scandals linked to planning permission and thus to the marketing of new homes. In international circles some compliance with local practice seems acceptable: in Mexico, for instance, international companies pay an illegal 'extra' charge to the post office to guarantee the proper delivery of their mail. Gifts and money can be involved here.
- **Pricing:** Often, we see prices stated as 'from' a particular amount. If a store advertises this, and only one product is for sale at such a price, is this unethical? Certainly it can annoy customers.
- **Images:** Advertising that offends or presents an unsuitable stereotype may be best avoided. Many people will remember the problems with Benetton's startling images of people dying of AIDS or on death row, and campaigns involving children or elderly people can present difficulties. Sexual images, too, are permanently in the news and, while sex does seem to sell, too much in this direction can prevent some people from buying and cause resentment among others.
- **Clash with culture:** Nestlé ran into problems promoting baby milk in the developing world. Advertising was felt to put people off breastfeeding and the cost meant that some mothers diluted the feed and caused nutritional problems for their babies.

There are many issues here: deception in selling, industrial espionage, wasteful packaging, confusion pricing, clashes with the environment and more. And it would be wrong to leave this topic without mentioning tobacco and alcohol. Is advertising some products simply 'wrong'? It is not illegal (though there are certain restrictions), but it attracts much criticism. Without going into the details – and it is a confused picture – it makes the point that marketers always need to exercise responsibility, and that judging the mood of the market on such things is very important.

What does all this mean? It is another given, another area that marketers must take into account in deciding how they do things. It is not always easy to judge; certainly breaches of advertising codes are sometimes a case of sailing too close to the wind and the line is not easy to interpret (except perhaps with hindsight). The effects of getting it wrong are clear.

- *Illegal action* will result in legal sanctions.
- *Breaching industry and other codes* can result in sanctions
- *Any action that consumers disapprove of* can result in bad publicity, reduced sales or worse – such as a boycott.

The reverse is also true. Organizations that get this right, certainly those making striking attempts to be ethical, can attract good publicity and see sales rise as a result. For example, many products are now promoted to stress 'green' benefits (though, if such claims are spurious, this can do more harm than good). A growing number of organizations now use an overt embracing of manifestly ethical approaches as a main platform of their offering. Certain banks and investment funds are just one example. In the many ways that this area impinges on consumer attitudes, it is an important one for marketers to work at. Besides, as has been said, one would hope that those involved in business would want to feel that what they do causes no problems of this sort, indeed contributes positively where possible, too.

## 2.4   Market segmentation

Another factor that exists and can help or hinder the marketer is the concept of market segmentation. Markets are not one large amorphous whole, everybody who buys a particular product: they divide into parts – *market segments*. These are parts of a larger market that may be selected as a discrete target and in which potential customers have a recognizable profile.

### 2.4.1   What is a market?

The market is a large collection of people somewhere in the world, the very people whom companies have to inform about what they can offer them. This may be in the way of benefits and satisfactions to resolve their problems and needs. However, we must consider that a proportion, and possibly a large proportion, may have no interest in the product or service being marketed. In fact they are not really part of the company's target market.

Therefore, what the company must do is clearly identify those people, or organizations, that are actually interested in their products/services. Only once a company has done this can it really use the concept of marketing in such a way that it can influence potential markets. A market segment is essentially a group of people for whom the product and all its benefits are suitable.

Ideal market segments meet the following four criteria, they must be:

1. **Homogeneous (similar) within:** Customers within the segment should be as similar as possible with respect to their likely responses to the marketing-mix variables and their segmenting dimensions.
2. **Heterogeneous (different) between:** Here the customers, within different segments, should be as different as possible with respect to their likely responses to the marketing-mix variables and their segmenting dimensions.
3. **Substantial:** Here the segment should be large enough to sustain the costs of accessing it and be profitable.
4. **Accessible:** It must be possible, and cost-effective, to communicate with the group that constitutes the segment. It is perfectly possible to identify very small segments that fail to be useful in marketing terms because there is no good way to access them.

There are many examples of incorrectly defined businesses. Let us consider the vast leisure industry. Sports shops selling running shoes, tracksuits and associated athletic wear have been known to define their business in terms of providing athletic apparel and try to aim their business at athletes. Yet much of the apparel sold by this type of shop is worn not by athletes, but by people who wish to be fashionable, or simply have good general leisure wear. Even an apparently simple market consists of segments, in this case at least two: athletes and a more general sector, and probably more.

We can see, therefore, that it is important for organizations to identify correctly the business they are in, and relate it to the characteristics of the market. Thus, they can correctly identify their potential customers. This will ensure that all their marketing efforts can be directed at potential customers so that resources such as time, money and effort are used efficiently. By carefully considering the makeup and characteristics of the potential markets there is a greater chance that the organization will be successful.

While on the subject of markets, it is worth noting one point: markets may be small and local or worldwide. The terminology here defines a number of aspects of this.

- *Exporting* tends to imply that goods – or services (so-called *invisible exports*) – are marketed and delivered from the country doing the exporting to other countries.
- *International marketing*, on the other hand, implies that the company itself is international and that its overseas business involves a greater presence overseas. This can, in turn, imply joint ventures, subsidiary companies and overseas offices.

That said, whatever the market, the way it must be approached is similar in principle.

### 2.4.2   Market segmentation

The concept of segmentation opens up a number of different ways, and scales of action, for marketing approaches. Key here are the following.

1.  **Mass marketing:** Here all buyers are considered as the same. This has the advantage of economies of scale and reducing costs. However, the concept of 'one size fits all' is difficult to sustain in today's drive for mass customization and differentiation. Many organizations are now moving away from mass marketing. Even Coca-Cola, which was originally available in only one size of bottle, is now available in a number of types and sizes of container as well as a number of different formulations, spreading its appeal into different sectors. The earlier concept, that of offering only one product, is not now in keeping with the requirements of the marketplace.

2.  **Segment marketing:** Normally this is considered to concern a *large* identifiable group within a market that has some attribute that makes it different from the mass. The actual differences may not be that great. However, they are discernible and therefore the consumer can have a product or service that more closely meets their needs.

3.  **Niche marketing:** Niche marketing is seen as applying to a more narrowly defined group, typically a small market whose needs are not being well served. Often, these groups will pay a premium price for the benefit received.

4. **Local marketing:** Here, the benefits being offered reflect the character of the region or small location being served. This could be a small country town or village where there is only one of a certain type of business.

5. **Individual marketing** (also known as *one-to-one marketing* and *micro-marketing*): This offers customized products or services to meet an individual potential customer's exact needs. This type of marketing approach is not uncommon in the automotive industry. Certainly, Volvo can customize a car from their range with the choice of about four thousand options of colour, engine, trim and so on (and other motor manufacturers offer similar approaches).

While one methodology, for instance a focus on demographics, geography or the behaviour of consumers, may predominate, none of these methods can be considered as unique. It is possible to consider a number of niche markets within a segment. Also, possible individual markets may exist within a niche, such as in the example above of Volvo. This is especially true in business-to-business markets, although the concept has been extended to a number of consumer markets.

### 2.4.3   Advantages of segmentation

Obviously, segmenting markets requires thought and analysis. This uses the organization's resources, so the work must be financially justified. The benefits of undertaking such work must therefore be identifiable and shown to be of value.

- **Ability to compare marketing opportunities:** The major benefits can be considered as being able to see and compare marketing opportunities that exist within the marketplace. These could be in the form of gaps in the market, where a particular benefit is not readily available from other suppliers. Therefore, the attributes required could be offered to that particular segment to fill the perceived gap. Care must be taken to ensure that the segment is sufficiently substantial to allow an adequate return on the investment.
- **Effective allocation of marketing budget:** Targeting segments can also help guide the effective allocation of the marketing budget. The aim is to concentrate expenditure on markets that will provide the highest return, and hence be the most profitable. It helps to ensure that the marketing effort is not wasted on products and services that cannot be effectively, or competitively, offered by the organization.
- **Ability to make adjustments:** Segmentation also allows an organization to make fine adjustments to the marketing mix specifically to suit the market. It may be necessary to modify only one aspect of the mix to change the effect or perception of the offering, and therefore make it more appealing to a chosen segment.

The concept of segmenting markets should ensure that marketing can achieve the specified objectives of the organization. This can be done only if:

- the characteristics of the individual market are known;
- the influence of specific buying groups upon those markets is understood;
- promotional activity is directed to the specific market segments;
- these segments are exploited to achieve the defined marketing objectives.

The fact that markets exhibit this split into smaller segments makes marketing much better able to select its approaches in a way that is most likely to maximize the profit potential that is sought.

## 2.5   Summary

Every factor of the sort reviewed here is something of which marketers must be aware. They must keep an eye on them, even attempt to predict changes coming, and work actively to ensure their approach to marketing recognizes the actual realities of the world in which they operate. With these factors in mind:

- restrictions can be recognized, and maybe their impact can be avoided or reduced;
- given creative consideration, examination of such areas is one of the ways in which opportunities can be spotted.

The keynote of everything here is that it should be dynamic. The marketing job is not just to accept and take on board external factors that will affect the business: rather it is to monitor a changing canvas and predict and react in a way that will help marketing to succeed. From this first part of the book, therefore, a number of further things should now be clear.

- Marketing is a complex process (though the individual elements of it are manageable, the orchestration of it is certainly complex).
- Marketing must work with the realities of its situation and deal with those things, primarily dealt with in this chapter, that exist and that are bound to affect how it can work.
- Marketing can function only if markets are understood (and this may demand research and the analysis of the information it produces).
- The product and price must be well chosen – competition gives no quarter to ill-conceived ventures.
- Given all that has been said so far, marketing is likely to proceed most effectively as the result of some creative and systematic thought, analysis and the formulation of a clear plan of action.

Such issues as research, pricing and compiling a marketing plan, indicated here, are all reviewed in more detail later. Next, we turn specifically to the first P of marketing: product.

## *Part II*
# Fundamental issues

With an overview of the broad marketing process in mind, we can now begin to consider specific areas of marketing decision and action that need attention ahead of communicating with the marketplace. With a product (or service) in mind, designed – engineered – and in production, how the characteristics of that affect its marketing must be considered. A price must be set for it. It must be decided what route it will take to market – which channels will be used. And all this must be based on sound judgement, with that very often having to be underpinned by research to reduce risk.

In this part, the disparate topics of the four chapters continue to set the scene and explain the wide span of activities encompassed by marketing.

First, we review the product, discovering that this is a complex issue where understanding is necessary of how a product can be viewed and dealt with in order to maximize its marketability. Second, we will see how the price on a product is an integral part of its marketing rather than simply a costing exercise, and that how price is set and used contributes to marketing's potential success. Third, we explore how – while marketing is certainly not a science in the sense that if certain rules are applied the outcomes are guaranteed – risk can be reduced by making decisions on a basis of sound data obtained through research. And, fourth, distribution is defined and shown to be as much a variable as any other aspect of marketing; decisions here too affect marketing's chance of success.

Additionally, the creative aspects of marketing begin to be made clear throughout this part of the book: marketing is as much about finding new ways to do things as deploying tried and tested methodology.

*Chapter 3*

# Product considerations

Simplistically, marketing is designed to move products from the producer to the consumer in a way that makes a profit. As we have seen in Part I, this is a complex process and many disparate activities are involved. All are dependent on the product. What is the 'product'? Well, for a start, it may be a service. Furthermore, exactly what it is directs the kinds of marketing approach needed. If the product is canned soup, then marketing is directed at housewives, except that this is an outdated concept and perhaps we should say 'homemaker', because many different people buy soup, including a lot of men; so too do many different kinds of establishment in the catering industry. The job of marketing to customers as varied as a housewife or a hotel chain is clearly very different. Compare this with Boeing selling jet airliners to airlines. Just two examples, but the way this must happen is chalk and cheese. Every product demands its own specific kind of marketing, and, even in an apparently discrete area such as engineering the range is broad.

Here we review the range of 'products' and brands with which marketing is involved, to see how different approaches are involved and something of what choices must be made to address them. *Brand* is the name applied to a product selling under a distinctive brand name. It both applies to an individual product, such as a Mars bar, and is an umbrella term describing a range of products that may be linked (so, here, including Mars ice cream). Brands can be powerful and worldwide, as are Sony and JCB.

## 3.1 Horses for courses

### 3.1.1 Major marketing sectors

The nature of the product and of the kind of organization involved directly affects the way marketing activity is planned and deployed. The next few sections illustrate the range of possibilities.

### 3.1.1.1    Consumer goods

First we will consider the marketing of what are called *consumer goods*. Let us be clear first what this terminology means. Consumers are customers in the sense of the public – you and I, buying for our own use and doing so from shops or other outlets to which we have access. The core area of consumer products is referred to as *FMCGs*. This stands for *fast-moving consumer goods*, those things that turn over fast because consumers want to make routine purchases (buying household products such as soap and toothpaste often).

Consumer product marketing is perhaps the most high-profile end of marketing. It is what many people think of when the word *marketing* is mentioned. It is characterized by being:

- very visible – these are products seen everywhere;
- directed at large markets – everyone is a customer for toothpaste (well, except maybe the guy who sits next to you on the bus);
- promoted in many media – be that television advertising or advertisements in glossy magazines;
- backed by large budgets, especially for promotions – which are necessary to reach the large markets and do so repeatedly;
- highly competitive – in many product areas there are many companies making essentially very similar products (though, as we will see progressively, small differences do matter in the market);
- reliant on the creation of a brand image (that is the product or service, the name and whole 'personality' that goes with it);
- creative in approach – witness the style of many consumer advertisements (though these may well be originated by the advertising agencies much used in this sector to plan and organize the promotional campaigns on behalf of their principals).

In consumer marketing many of the large firms have a string of brand names – and all need marketing. Sometimes this is apparent, as with the many products marketed under, say, the Nestlé name. Other companies use a list of different brand names and the association with the main company is not featured strongly (as with Lever Brothers). Multiple brands and wide product ranges give rise to the job of *product manager* (also sometimes called *brand manager*, this refers essentially to people who act as 'mini-marketing managers' for a single brand rather than a whole organization).

Given the scale of marketing in this field, these are big jobs, and the task of achieving the planned market share is a challenge. Market share is a classic marketing measurement stating the amount of a product sold as a percentage of total sales of similar products (it implies that the total sales of the particular product, instant coffee or whatever, are known).

So, those people marketing consumer goods are certainly in the heartland of the marketing world. Opinions differ. Some people feel that this is the only *real* marketing. Certainly, it is a sophisticated form, one that utilizes the full panoply of marketing techniques, though the fact is that the competitive pressures and the challenge of making anything successful in the marketplace are just as real whatever the product.

Engineering may, of course, be essential to the development and production of such products.

### 3.1.1.2   Industrial products

This phrase tends to characterize a very different area of marketing: that of 'heavy' goods sold to industry rather than to Joe Public. Again, the range is considerable. It includes a considerable involvement with engineering. The product might be machine tools or products necessary to them (from spare parts to specialized oil), as well as a whole range of components: items bought to become part of whatever product the purchasing organization makes and sells. It includes complex items such as ships or space shuttles. It includes a mass of products necessitated by what is called *derived demand* (this phrase describes the way in which the sale of one product follows from the sale of another (see Panel 3.1).

### Panel 3.1   *Derived demand*

In a situation where derived demand exists, as for example in a company manufacturing and selling glass bottles, these might be bought by a brewery and filled with beer. Thereafter, the success of the beer in the market dictates how many bottles are sold. The design of the bottle may be such that it helps produce an attractive image and sell the beer, in which case the bottle manufacturer may need to become more involved in the business of selling what is put in its bottles than might at first be thought necessary. Such involvement is a routine part of marketing activity in some fields.

There is some variety here, but industrial product marketing has a number of important characteristics.

- There is an inherently smaller number of potential customers – everyone may need toothpaste, but I for one do not have or want an industrial lathe in the back bedroom.
- It often involves long lead times, since products are designed and engineered – a new car may take four or five years to produce, but a new airliner twice that time.
- It addresses 'professional buyers' – often people paid to buy and trained to get the deal they want.
- People working in it perhaps need a technical background, qualification or understanding. This will vary of course but can be extreme (the lowest form of life in some marketing departments has a PhD in nuclear physics). There are plenty of opportunities in marketing for those with specialist technical knowledge or qualifications.
- More specialist and targeted approaches are needed. It is wasteful to advertise, say, heat exchangers on television, but advertisements in technical journals still need to do an effective job.

- Personal selling is likely to have an important role – the final link in the chain is often a personal contact.

Marketing is just as necessary in these areas and may well demand a greater amount of technical expertise alongside it.

*Note:* industrial marketing and the next category are not precisely separated. There is an overlap.

### 3.1.1.3   Business-to-business products

These have much in common with industrial products. The term *business-to-business*, which is how industrial products are sold, is self-apparent and came into use as a phrase more recently than *industrial marketing*. The difference is primarily that this category omits the heavy end of industrial marketing.

So products here are those bought by offices and factories and by subgroups such as what has recently become known as *the SOHO market* (the letters stand for *small office, home office*). They include a vast range of things. To illustrate, the following is a pure miscellany by way of example: telephones (and telephone systems), office furniture, paperclips, computer disks, technical journals, stationery, business books and cupboards (on the basis that even the business that has everything needs somewhere to keep it!). There is a further overlap with computers, software and other high-tech product areas mentioned later in this chapter. Essentially, here are all the products a business must buy to keep itself and its people operating.

Brands can be as important here as in consumer markets. Indeed, some brand names appear directed at both (you may see a Mercedes car as an attractive prestige product and then find a dirty delivery van outside your door with the same logo on its bonnet).

But there is another category that, again despite an overlap with what has been mentioned already, is important – not all products are products.

### 3.1.1.4   Services

*S*ome 'products' are services. Indeed, please take it throughout this book that the word *product* may usually be taken as meaning 'product or service'. These may be sold to the consumer sector, as with, for instance, dry cleaning, tax-free savings accounts and film processing, or to business and industry as with industrial design, contract ploughing and staff training. And they can be sold to both, as with accountancy, insurance and travel. As we have seen, certain engineering activities fall firmly into this category. Again you will find a moment's thought shows that there are many examples.

How is service marketing different? Services:

- are intangible – the fact that they cannot be tested by potential customers in advance of purchase in the same way as a product can certainly makes for a different approach to marketing and selling;
- are inextricably bound up with service – they are the 'people businesses', and marketing and the organization of delivery of the service overlap;

- interface very directly with customers – much more closely than in some other businesses;
- allow change and flexibility to be greater, and sometimes easier and faster, than in other kinds of business (producing a new insurance policy, say, is inherently easier than producing a new jet fighter).

The immediacy of services, in customer service for instance, must be reflected in the way that their marketing is organized. It is also worth bearing in mind the way in which services have grown in importance in recent years. This is especially so in countries such as the United Kingdom, which have seen their manufacturing base decline.

Having defined things to this extent, and with some overlap continuing, we can look at two further 'sectors' where marketing must be implemented in ways that reflect their special nature (and then, briefly, at a number of specialist industries).

## 3.2   Specialist areas of marketing

### 3.2.1   Social marketing

This is a jargon term describing a specialist area of marketing activity, one that has become a major force in the marketing world: that of marketing activity in organizations where profit making is not the objective. A charity, for instance, uses marketing communication to raise funds and ultimately to allow its philanthropic work to proceed.

Marketing used traditionally to describe effort designed to produce profit. But not every organization is profit making. Well, some, it must be said, fail to make a profit despite their best efforts, but here I am identifying those who do not *want* to make a profit. It should be remembered that only the tiniest amount of money makes the difference between breaking even and making a profit or loss; in other words achieving whatever financial outturn may be required needs some skill. Three main sectors come to mind here:

#### 3.2.1.1   Charities

These days many charities are, by any definition, big business. Their target market is contacted to produce funds, and marketing methods may be used in different ways (to change public or government attitudes, for instance), but marketing is real and important for them and they need marketing talent to achieve their aims and fulfil their charitable purposes. Such organizations are an interesting option for some of those wanting to pursue a career in marketing.

#### 3.2.1.2   Government

Both local and national government has marketing operations. These may be on a grand scale, as with the advertising undertaken to highlight the dangers of drinking and driving or the need to adjust to self-assessment taxation systems. Or they may be less major and more local, as with local authority schemes to help small business.

Sometimes the target of such marketing is more bizarre. In Singapore, for example, government messages are much in evidence. At one time, when population increase was seen as being desirable, television advertising called for people to fall in love. I contemplated writing in to say my visit was only four or five days and I would do my best, but felt the relevant ministry would probably not see the funny side of it. Nevertheless, there are significant and interesting opportunities here for some in marketing.

### 3.2.1.3   Quasi-government and others

The above category overlaps into a whole range of other bodies: government agencies, trade organizations (such as the Wool Marketing Board), educational establishments and professional bodies (such as the Institute of Chartered Accountants, who promote the merits of working only with an accountant with the appropriate qualifications, and similar bodies in engineering, including the Institution of Engineering and Technology). There exists a section of bodies here that have a worldwide remit. Again, this sort of marketing is much in evidence to us all.

Another sector also exhibits specialist characteristics.

## 3.3   Marketing services

A plethora of specialist services exist in marketing. Advertising agencies create advertising for their clients and are specialists in selecting appropriate media. Subsectors of this are, for example, agencies specializing in sales or point-of-sale promotion. Market-research agencies conduct surveys to identify markets, test products and try to reduce the risk inherent in the marketing process.

Even more specialist agencies include those concerned with packaging design, photography and copywriting. As well as working as a part of the marketing process, all such agencies have to market their own services too, so there is a need to deploy specialist skills in that regard also.

Beyond all this there is the question of industry specialization.

## 3.4   Industry specialization

Of course, all industries differ to some extent. Some industries are very specialized (not so much technically, but in ways that affect their marketing and the people who undertake it). The following are mentioned by way of example; obviously it would be possible to consider many more, but this much establishes the range. Such include high-tech and information technology, professional services, pharmaceuticals, financial services and more. Engineering features here, too – certainly some engineers work in professional services.

There is no space here to review every industry. Many have particular, and sometimes topical, characteristics. Industry is not the only differentiating feature here, however: the size of an organization and the resources it can therefore bring to bear also affect how marketing is carried out.

It may be illuminating to think about one area of which you have some real knowledge. This might be the area in which you currently work or have worked previously, or one of which you have some other experience. Think about how the nature of the business stems in turn from the nature of its market and customers, and how the nature of the products or services themselves influences the way its marketing is executed.

It is also something you can consider to a degree when an advertisement hits your consciousness when you are watching television or reading a magazine.

## 3.5    The product in the market

Even when the product is tangible and easily defined, there are other matters that necessarily affect the precise form that marketing takes. *Market* is a broad description. In the last chapter market segmentation was discussed. Focus on a market segment, or its smaller cousin the niche market, helps make marketing decisions easier. Instead of being directed broadly, marketing can then concentrate, more specifically, on one particular sector and tailor its approach to that. Thus, a pharmaceutical product may aim to appeal only to farmers raising beef cattle and the vets they call on to secure the health of their livestock; and a specialist oil may be suitable for use only in a narrow range of machines.

At this juncture a key point should be noted: in whatever field a product is, it will succeed only if it is 'good'. No one can successfully market a car, let us say, that fails to start two out of three times, does only two miles to the gallon when it does go and rusts within the week. This is an exaggeration, of course, but the point is that a product, whatever it is, must perform, and perform against competition. The word *good* was in inverted commas earlier to make the point that to be successful a product does not have to be the best of its sort – ever – but it does need to provide value for money and meet the needs of a sufficient number of customers to pay its way. And some low-cost products are successful. No one would pretend that the Bic throwaway ballpoint pen is in the same league as a Mont Blanc, but they sell in their millions and most customers are satisfied that they get value for money.

So a 'good' product is a prerequisite of marketing success. It is sometimes said that marketing is marketing products that don't come back to customers that do.

Clarity of view about the product and its relationship with customers is key.

### 3.5.1    The product USP

More jargon. Although we said above that any product must be good, this is insufficient in isolation. A product must be good compared with its competitors, and that is what marketers refer to as having a *USP*.

The USP (this stands for unique selling proposition) is whatever it is about a product that differentiates it from others and makes it appeal to customers. The term *value proposition* is also used in this context.

There is a difference here between being the best and having a USP that is effective. *Best* implies superior to, and thus perhaps more expensive than, other products; it implies things that are put in rank order. Here we are viewing things more like Darwin's survival of the fittest: the USP gives a product an edge in its chosen sector. Thus, it is as necessary that a simple, less expensive product will appeal to the market as a sophisticated one. Somewhat like evolution, marketing is involved in a constant search for new elements that will keep a product ahead. Unlike evolution, which proceeds by chance with no preconceived aim, marketing *intends* to succeed at this process.

The USP is as applicable to image, and to communications elements such as advertising, as it is to tangible factors about the product. At the end of the day it is usually a mix of disparate factors that contribute to a product's success in this way as one overriding factor.

The next subsection comments on a systematic way of viewing products designed to help with this process.

### 3.5.2    The augmented product

*The product* describes more than just the product in marketing terms. Let me explain. The product, what might be the core product, is not enough in itself to guarantee marketing success. It must be added to, to create an entity – the *augmented product* – that allows marketers to make more of it.

Three layers, as it were, are involved here, and the more a product is in competition with others the more the outer layers must be worked on:

- The **core layer** is simply the product itself, with its price inherently tied up with it.
- **Layer two** adds elements to the product designed to make it more appealing. So here we might include packaging, a brand name, a level of quality, design or style features, and all the things the product does or means (the benefits, of which more later) to actual and potential customers.
- **Layer three** adds more, and adds things that might be described as less inherently part of the product itself. Such things include warranty arrangements, delivery (and perhaps installation) arrangements, training in use, after-sales service and financing.

Some of these factors just add to a product's appeal in the way that an attractive container can, in part, influence what is chosen (to match the décor in a bathroom, perhaps, for a consumer product such as a shampoo). They can also become a significant part of the product package. For example, hire-purchase and leasing arrangements are almost as important as the car in that market. Such things influence both what is bought (a Ford or a Mazda, say) and where it is bought from (two different distributors will deliver identical cars, but arrangements such as finance may vary considerably).

This helps explain just how important the product is as a marketing variable – one of the three key Ps of the marketing mix (product, price and presentation). It is not in any sense fixed and there are many decisions to be made and many tasks to be undertaken to create a product package that is likely to thrive in a competitive market.

The situation surrounding such marketing work is, of course, dynamic. Things change: when one competitor amends its product package to help differentiate from competition, then its competitor must react and change too to stay ahead. Often minor changes between products are significant. To illustrate this, consider a product such as soap. You may not like the smell of some, but probably there are many others where frankly there is not much to help you choose between them. Price is important, of course, but the same is no doubt true of brands selling at very similar price. So you, just like other customers, find some detail that you allow to prompt the purchase. This could be a special offer (two for the price of one, or some such), but it might be minor product details too. For example (how sad is this?), I like Pears soap, in part, because the indented surface allows you to stick the remnants of the old worn-away bar to the new one and avoid waste.

### 3.5.3   Product positioning

Another tool of marketing that needs consideration is *product positioning*. This term describes the marketing tactic of positioning an individual product with precision within the spectrum of other broadly similar competitive products so as to achieve maximum focus and marketing effectiveness.

One example will make this process clear. Consider motorcars. The range of makes and models available ranges from city runabouts to luxury saloons and sports cars from makers such as Rolls-Royce and Porsche. Yet all cars get you from A to B. In introducing a car, a manufacturer must be clear how it will be positioned in a way that involves both product and price. The complexities here are considerable, but, to continue the example, are the quality, design and price of the car going to put it at the 'top' of the market, or is it going to be aimed at people wanting only basic transport at an economy price? This is investigated in more detail later.

## 3.6   Summary

Marketing approaches are many and varied. The marketing job is, as much as any-thing, concerned with deciding which of many possible things that *might* be done *will* be done and how various things will be fitted together (the promotional mix). What is likely to suit, and work best, is primarily influenced by two things:

- the product or service
- the market and customers towards whom it is directed

All marketing activity must reflect the nature of the world in which it operates. What is right to do in marketing toothpaste may be very different from what is needed to market ball bearings. The difference here is simply interesting. But what is crucial in marketing is:

- that what is to be marketed should be well chosen – a poor product will rarely, if ever, sell more than once;
- that marketing activity should match the product in every possible way.

The product and its price are inherently associated (as the reference to product positioning made clear above), so it is to price that we logically turn next.

*Chapter 4*

# Pricing policy and tactics

Unless the product is to be given away, it must have a price. Setting a price is more than picking a 'suitable' figure out of the air: it is an important element of marketing and needs serious and systematic consideration not least of customer attitudes to price. The tactical use of price is also very much a part of ongoing marketing activity. This chapter reviews the way things work with regard to price.

## 4.1   How much?

Every product has a price. The problem is deciding what it should be. It affects not only profit – clearly – but also image. The word that goes most easily with *cheap* is *nasty*, yet everyone wants a bargain or, since a bargain is essentially something worth more than it costs (and therefore rare), what they actually want is *value for money*.

Price puts out many messages. It can say something is classy, of good quality, fashionable or shoddy. There is an apocryphal story that one of the first astronauts to go to the moon was asked what he thought about in the last few seconds of the historic countdown. He thought for a moment and said, 'I remembered that there were five hundred thousand working parts in the machine underneath me, and that, in every case, the contract had gone to the lowest bidder'. There are situations where low price does not boost customer confidence. Who wants the cheapest insurance, stockbroker or (if you go private) surgeon?

Price has other psychological impacts. For example, all the research shows that if a price is just below a round figure, £9.99 or £19.99, rather than £10 or £20, people will buy more. Probably no one really understands why, but many manufacturers and retailers use the fact and price accordingly. (The other reason this is done is for security – at such prices the shop assistant must give change, and thus must open the cash till and record the sale. Sadly, staff pilferage is a significant cost to be avoided.)

The same applies to higher prices, with figures set just below £500, £50,000 and so on.

Similarly, the thinking about discount levels for retailers or wholesalers, quantity terms and so on all have to take account of the way people think about the figures. The use of price at annual or seasonal retail sales is another example. Of course, everyone wants to save money, but just how it is put is significant. Which sounds better on a £100 item: 'Half-price' or 'Save £50'? And, if it was reduced to £55, how different does 'Almost half-price' or 'Nearly £50 saved' sound? The ways of wording this kind of thing are, of course, legion.

Remember that price does not only inform customers of the cost to be paid, but is also an indicator of quality. Faced with a choice, especially of reasonably technical products such as cameras or music systems, it is difficult to make a logical decision between them. Price offers a measure – the more expensive one is surely of better quality. With some products, for instance fashion brands, this is taken to extremes: the high price does not really buy extremely high quality compared with other options, but it buys status – people know from the label that a premium price has been paid. Some brands play on this. A company like Sony does offer good quality, but is resolutely higher in price than some of its competitors. High price may imply much more. Long-term service is one. The Canadian outdoor-clothes company Tilley offers lifetime guarantees on its famous hats, for instance. Again, such principles apply right across product categories.

So, setting pricing policy is a vital part of the marketing people's jobs. And it is not a once-and-for-all decision: *we will charge this much, and put it up by a percentage for inflation each year.* Price can be, and is, used tactically, to steal an edge on competition.

## 4.2   Setting price

There are many factors to be borne in mind here. Here are some of them.

- **Cost:** Obviously, things must be sold for more than the cost of making and selling them, at least if a profit is required.
- **Tax:** Where tax will be added and condition the price that customers see and pay, this is a real consideration, especially with products carrying high taxation (such as cigarettes and petrol).
- **Value:** This is obvious, but also relates to circumstances: an iced drink on a hot day may command a premium, especially in the only café for miles around.
- **Market conditions:** Especially whether demand for a product is increasing or decreasing (something that may happen rapidly or slowly).
- **Geography:** For example, in some places, products reflect the cost of isolation (and distribution costs).
- **Legal restrictions:** Laws may forbid certain price tactics (such as agreeing prices with competitors).

- **Consumer preferences:** The more desirable something is, the higher a price may be paid for it – for example, a new 'must-have' gadget such as an iPod or, at the other end of the scale, a work of art.
- **Price-sensitivity:** Some customers are more price-sensitive than others. A pensioner on a limited budget, for instance, may forgo buying something completely if the price rises even a small amount. A factory may delay replacing machinery.

So, with all this in mind, how are prices set? There are four basic approaches to pricing:

- cost-based pricing
- market-demand-based pricing
- competition-based pricing
- market-based pricing

It is well worth determining price levels in a way that combines elements of all four approaches. Let us look at them in turn.

### 4.2.1   Cost-based pricing (the accountant's approach)

This is the approach similar to the way an accountant would calculate the price for a product. It is based on total cost of product, including production and marketing costs, plus an allocation for overheads plus the target percentage to provide a profit margin. The total gives a selling price.

It is not without its problems. Cost calculation is based on a predetermined level of demand and production. As these fluctuate, so does the product cost. It ignores market factors such as demand and competitors' actions. The way overhead cost allocation is done can lead to a wrong pricing decision. However, a major benefit of this approach is that it can help indicate *minimum* price levels.

### 4.2.2   Market-demand-based pricing (the economist's approach)

The aim of this approach is to explore the effect that different prices may have on the demand in the market for a product. Here the marketer will try to calculate the break-even point (produced by varying volume forecasts) based on different selling prices. This approach brings into focus the impact of price on volume and tries to find the most profitable price–volume ratio. Marketers must ask themselves how many units of a given product they could sell at different price levels.

The *advantage* of the market-demand approach is that it brings together price calculations with market-demand realities; that is, if demand for a product tends to be a function of its price, then this should be a determining factor in the decision.

The *disadvantage* is the difficulty of estimating the effect that price variations may have on product demand: one has to estimate how much one can sell in units for a given price level. Given this problem, an easy way to establish price elasticity is to examine the historical performance of similar products at a number of different price levels to study the effect on sales of price.

### 4.2.3    Competition-based pricing

The method used in this pricing approach considers the prices set by competitors in the marketplace. Competitive pricing can be approached in a number of ways:

- prices can be set *above* those of competitors' products;
- prices can be set *below* those of competitors' products;
- prices can be set *at the same level as* those of competitors.

Note that legislation in some markets may prevent collusion between competitive firms being used to fix prices and stultify competition (it may also regulate other specialist areas such as electricity supplies in UK).

### 4.2.4    Market-based pricing

In this approach prices are based on the value satisfaction that the product delivers to the buyer, or, rather, the *perceived* value. This perceived value can be a result of:

- value for money influenced by all aspects of the organization, and its product and service;
- image affected by status (endorsement by opinion leaders, exclusivity or promotion);
- reflection of different and distinctive market segments putting different value on a product's performance;
- price barriers that are apparent in different segments.

The key to market pricing is to make an accurate assessment of a market's perception of the value of the product. Market research may be needed to avoid two dangers:

- **overpricing** because of an inflated view of the value of a product: almost inevitably the perceived price will be a qualitative judgement made by the buyer relative to their experience of the competition;
- **underestimating** the real value and charging less than is possible, thus reducing profit earned.

Whatever is done, there are clearly a considerable number of, sometimes conflicting, issues to be borne in mind. Price setting may very often be a compromise.

## 4.3    Pricing strategies

Given a price range in a particular area of the market, a decision has to be made as to where in that range to locate a particular product's price. This is a strategic decision made on the basis of corporate and company objectives, which may include some or all of the following – the need to:

- achieve target return on investment or sales;
- stabilize prices;

- maintain or improve market share;
- meet or prevent competition;
- maximize profits.

Let us now look at some of the most important tactical approaches to price.

### 4.3.1　Skimming price policy

This strategy sets a price at the top of the acceptable price range. Among its *advantages* are when it is used:

- on a new product in the early stages of its life cycle to recoup high investment;
- to segment the market;
- to prevent pricing mistakes by having prices too low (it is easier to reduce prices than to increase them if the wrong price level is chosen);
- to limit off-take if planned capacity or stocks are not adequate.

It can have *disadvantages* too:

- it may attract competition;
- low volume may not suit production objectives;
- it may mean that consumer awareness and acceptance will be slower in the introductory product life stage for a new product;
- it may be more vulnerable to less advantageous economic conditions.

### 4.3.2　Penetration price policy

This strategy is the opposite of skimming. A low price is set, often below the existing range of competition, with the objective of gaining maximum market penetration as quickly as possible; that is, low price, high volume.
　*Advantages* include:

- product economy of large-scale production;
- pre-empting competition;
- winning wide product allegiance for the future.

*Disadvantages* include the facts that:

- profit return is lower and payback period is longer than for a new product;
- it will be disastrous to profit if the product has a very short life-cycle;
- it can be difficult to overcome the psychological disadvantages of having to increase price if the initial price was set too low.

A good example of skimming pricing used to good effect that probably most people have noticed is with technology products such as computers. The new model is high in price for a while, but drops as further new models come into the market at high prices in their turn. The cycle is apparent to all, and some customers just wait a while.

### 4.3.3   Marginal cost pricing

In highly competitive situations one may have the opportunity of gaining business if a sufficiently low price is offered. The question arises, however: what is the lowest price to use at which it makes sense to take the business? One approach is to use marginal costing, which is defined as the cost of producing one more unit.

The cost of producing one more unit means that the existing sales volume is already covering the fixed costs, and then the costs of producing the extra unit are the variable costs. If a small profit is made per unit, then at least this is an additional contribution that would not have been there had we not obtained the extra business. It can be argued that, even at no profit, marginal business is worth having, as it may use resources that would otherwise stand idle. Generating this type of business, however, can ultimately eat into profits and depress the percentage of return on sales. The major use of marginal costing, therefore, is to answer the question 'Should I accept this order?' rather than as a pricing tool.

This type of strategy is more often used with price-elastic, high-volume products, where it is important to keep the sales volume up. That said, marginal costing is a common basis for export sales: to be competitive in foreign markets, elements of cost already covered by home sales are ignored in setting prices overseas.

A few guidelines might be useful in summarizing this important area of marketing.

- Pricing is not seen by the purchaser simply in terms of what is the cheapest, but rather as one element in the bundle of benefits; that is, what the image of the product is in the mind of the consumer/user and hence the perceived value.
- Customers never buy on price alone; witness the fact that the most successful operators rarely are at the low-price end of their markets. Yet too many salespeople will often argue that it is the sole purchasing motivator.
- Before marketers comment on competitive pricing strategies or price structure, they must make sure they have hard facts to work on. Otherwise they may well make pricing recommendations that are unnecessary or wrong.

### 4.3.4   Day-to-day tactics

A host of devices are deployed to boost sales and profits just by the way prices are set (perhaps only temporarily). The following examples give a feeling of the range of tactics involved:

- **Variable pricing:** Essentially, this means different prices at different times (as with trains and cinemas) or for different people (such as half-fares for children or special rates for pensioners).
- **Prestige pricing:** This sets a high price to encourage a feeling of exclusivity, in the way that caviar, inexpensive in local markets, is priced up when exported.
- **Set prices:** This is the retail tactic that shows itself as everything under £5 or whatever.
- **Bundling:** This sets a price for a 'package' of some sort such as a computer with a particular suite of software or a holiday with various add-ons such as car hire.
- **Trade-in allowances:** As with trading in a car or upgrading industrial machinery.

- **Quantity discount:** A host of pricing devices link buying more to a lower unit cost.
- **Seasonal discounts:** In retail often linked to seasonal sales or newly introduced products at a higher price – 'Just in!'.
- **Timing:** A price for today or this month only.
- **Price plus extras:** A basic price is made to look like good value, but other items need to be added on (for instance a digital camera almost never has any significant memory and this has to be bought as an extra).

Additionally, pricing may be affected by other elements of the product and purchase, as with the financial arrangements (loan or leasing) enabling the purchase of a car, say.

There is always a tendency to overreact in the face of what is seen as competitive price pressure. *Remember that brand leaders are rarely the cheapest. They are frequently among the most expensive.* If this seems to be telling us something, it is. If there is a likely danger in setting prices, it is more likely that setting prices too low does damage than is the reverse. High price literally finances good marketing and can create a positive cycle.

Pricing is of fundamental importance. It is an important, creative aspect of the marketing job. The marketing plan should include actual prices, costs and margins at various levels (i.e. the pricing structure).

Prices of major competitor products should also be included along with recommended prices for each year of a marketing plan, shown as actual prices and percentage change for each distribution (wholesaler, retailer, etc.) for each year. A brief statement of the pricing strategy and of the various pricing considerations must also be given to summarize the options and decisions. Panel 4.1 summarizes the price-setting process as a systematic approach.

*Panel 4.1   Six steps to price-setting*

First, the company carefully establishes its marketing objective(s), such as survival, current profit maximization, market-share leadership, or product-quality leadership.

Second, the company determines the demand schedule, which shows the probable quantity purchased per period at alternative price levels. The more inelastic the demand, the higher the company can set its price.

Third, the company estimates how its costs vary at different output levels and with different levels of accumulated production experience.

Fourth, the company examines competitors' prices as a basis for positioning its own price.

Fifth, the company selects one of the following pricing methods: cost-plus pricing; break-even analysis and target-profit pricing; perceived-value pricing; going-rate pricing; tender pricing.

Sixth, the company selects its final price, expressing it in the most effective psychological way, checking that it conforms to company pricing policies, and making sure it will prevail with distributors and dealers, the company's sales force, competitors, suppliers and government.

Companies can then apply a variety of price-modification strategies to the basic price (such as market-skimming or market-penetration strategies).

## 4.4   The role of price in creating image

With any product you care to mention, its price and image are inextricably linked. This is important because it has a direct bearing on how successful the product will be, and also, importantly, on how much profit it will generate.

So do customers want the cheapest product? Yes may be the obvious answer, but not if low price betrays inadequacies. What they really want is *value for money*. This means they are prepared to pay more for products of superior quality, and this applies whatever the product sold. Rolls-Royce and the word *cheap* just do not go together. The consumer expectation is that a Rolls-Royce will command a Rolls-Royce price, so much so that a virtue is made out of it when one refers to other products as 'the Rolls-Royce of X' – and charges accordingly. Many producers visibly use this in their offering. For example, the Belgian beer Stella Artois is sold as 'reassuringly expensive' and Caterpillar Tractors are high-priced with this equated to exceptional service.

How do we arrive at the right price? Well, the price must be reasonable. Now customers may well say, '*Your* "reasonable" and *my* "reasonable" may not be the same'. This is true of course, but within limits one can judge what the market will bear taking into account the segment of market and the level of competition. So now we have a framework for deciding whether there is a hole in the market that we could fill at a profit, and whether it is worth doing the necessary calculations to see if that would be desirable.

There is a further element that has entered pricing recently, and this is so-called *confusion pricing*. It is regrettably something that is happening more and more, especially with such things as mobile phones and their tariffs. The prices set are so complex that they are unable to be easily compared with those of the competition, so consumer choice is effectively limited. This may work for a while, but there is always a danger of a consumer backlash or a competitor setting themselves apart from what is happening in an industry and achieving positive differentiation.

Many techniques for using price exist, all designed to get better impact from the marketing power of pricing. 'Two for the price of one' is just one everyday example. Sometimes called BOGOF (buy one, get one free), it is something offered by many, sometimes to shift stock, sometimes as part of a plan linked to promotion.

## 4.5   Psychology of price and customer reaction to it

We saw earlier how price is important, but it is not the only consideration of the sale, so how do we make sure that we maximize the price without losing sales? Anyone knowing the answer to that one, every time, would be a millionaire. It does vary so much from product to product and to a certain extent the goalposts are always moving, as markets are dynamic.

For example, consider this. Having done all the research, you launch your product at a certain price on the market. To universal acclaim you find the market has an energy of its own and, because the take-up has been so good, you find that you are having difficulty keeping up with production. This is a classic situation and has happened many times in the past. You are now faced with the decision: do you raise the price to dampen down the market and maximize profit, or will this damage the consumer's expectation of the new product and perhaps start a price war?

Many situations that demand pricing decisions are more complex than this and demand wide review of all factors before a decision is made. However, a decision has to be made even if it is to do nothing and get round the problem another way. Your relationship with your new customers could be damaged, and a bad piece in the press could kill the product. We do not have to look far to find many such examples. Wrong pricing tactics can cause damage to the brand image, especially those brands at the top end of the market, and cause a price war that may be difficult to stop. There is no one as fickle as the customer.

Pricing is a complex issue, one that needs careful consideration from both the marketing end and the customer end. Get it right and marketing is likely to work better in every respect; but there are dangers and every aspect needs careful consideration.

## 4.6   A systematic approach

Ultimately, as with so much in marketing, the art/science elements of it must be combined and, although judgement is important, so too is a systematic approach.

A number of models of pricing have been considered but it is necessary to work systematically through a procedure actually to arrive at a price. Beyond what has now been said, the most common model is that developed by Oxenfeldt in his *Harvard Business Review* article entitled 'Multistage Approach to Pricing'. This model has six stages and, while not the latest, still provides a sensible and straightforward systematic approach to calculating the price that should be charged. This model is paraphrased in Table 4.1.

It is also worth noting an approach Hugh Davidson advocated in his excellent book *Even More Offensive Marketing*. This is similarly paraphrased here in Table 4.2.

Organizations are faced with a number of situations in deciding the most suitable pricing strategy. Often, the first problem to consider is how to price a new product. While this can be difficult when entering an existing market, it becomes more so if the organization is trying to enter a new market with its new product. The problem of trying to reposition a product usually requires a modification to the price to maintain or establish credibility in the new position. It is also sometimes necessary to consider a price change in response to threats from close competitors. Also, it will be necessary to modify the price as the product passes through the various stages of the life cycle.

*Table 4.1    Systemic pricing model*

| Stages | Description/action |
|---|---|
| Market opportunity analysis | It is necessary to select the specific market segments that the organization wishes to target. As we have already seen, different segments will expect different levels of price. Therefore, a clear policy on which market segments are to be targeted is crucial to ensure that an acceptable pricing strategy is developed. |
| Company image | The second stage is to consider the sort of image the organization wishes to portray, both for itself and the product or service in question. The image and reputation will have a direct effect on the price that can be charged and it must, of course, be compatible with the chosen segment or segments. |
| Marketing-mix strategy | It is then necessary to develop the marketing-mix strategy for what is being offered. The price is affected by and in turn affects all the other elements of the marketing mix. Therefore, all the variables must be considered together. The components must blend and portray a consistent image in the minds of the potential customers. |
| Pricing and the marketing plan | After the initial stages of this model have been worked through, it is then possible to start considering the most suitable overall pricing policy. This must, of course, fit with the overall marketing plan. |
| Develop and implement pricing strategy | The next stage is to develop and implement a pricing strategy. We have seen that prices are likely to change during the various stages in the product life cycle, so it is important that this be planned in advance. The other elements of the marketing mix will also change, so care must be taken to ensure that consistency of the overall image of the product or service is maintained. |
| Specific price for product or service | The final stage is to choose a specific price that will be charged for the benefits that are being offered to the market segment. Provided the groundwork has been undertaken then this price will be the market price and will result in the optimum sales for the product. |

Analysing the product and market situation is a good prerequisite to making pricing decisions. In doing so, various questions must be asked. Here are some of them.

*How large is the product market in terms of buying potential?*
**A thorough understanding and evaluation of the potential market are required before any consideration can be given to the pricing of the product.**

*What segments exist in the product market and which market target strategy is to be used?*
**We have seen that different market segments seek different benefits and hence expect to pay different prices. A clear identification of the segments and their profile is vital in price setting.**

*Table 4.2   Offensive pricing*

| Principles | Description/action |
| --- | --- |
| Know the price dynamics of the market | It is important to get a feel for what might influence price. Consideration of such factors as frequency of purchase, degree of necessity, unit price, degree of comparability and degree of fashion or status contributes to the price sensitivity of the company's product or service. |
| Choose price segments | Price-bracket segments of all markets. In general the strong brands are in the upper pricing segments while the commodity products and less known brands are in the lower segments. It is important that an organization can clearly define the segments in which it operates. |
| Achieve clarity of pricing | Potential customers must understand a company's pricing system. If it is not logical then they will not trust either the brand or the organization. Currently, many flout this rule – confusion pricing was mentioned – and UK rail ticketing is another good (poor?) example. |
| Always consider the alternatives | Pricing is often considered as a rather mechanical aspect of marketing, but it should be approached creatively. Price is only one part of the marketing mix. Therefore, the whole mix must be considered when analysing the pricing strategy. |
| Target price changes | Remember that price elasticity varies by type of consumer, shopping environment and occasion of use. Companies must make sure that they understand the way price elasticity works and use price promotions, not only to stimulate new sales but also to reward loyal customers. |
| Avoiding profit cannibalization when pricing new products | Companies should ensure that new products take profits from their competitors, and not from other products in their own product range. |
| Using pricing to optimise return on capacity | This is especially the case with perishable products such as fresh food. This will mean using all the available demand-forecasting tools, and having a good understanding of and being able to act upon price elasticities of different customer types. It will also mean analysing capacity utilization and using price to maximize it. A company must also make sure that its cost allocations are efficient. |
| Pricing mistakes/errors | If a company makes a mistake on pricing, it must admit it and remedy it fast. It is easy to make mistakes. Normally they do not really matter as long as they are spotted and efficiently and effectively adjusted. |

*How sensitive is demand in any segment to change in price?*

**This relates to price elasticity, as we saw earlier. It is very easy to get this wrong and misjudge the acceptable price range. Prices set outside the acceptable range of price levels in the potential buyer's perception will result in the loss of sales.**

*How important are non-price factors such as features and locations?*

**This is an area often missed by organizations when considering price levels. Products must be differentiated and it may often be better to enhance the product or service features as a competitive tool rather than use price.**

*What are the estimated sales at different price levels?*

**Levels of promotion will also affect sales, so this and other factors must be taken into account when looking at setting price levels.**

Finally, let me highlight some of the most common mistakes made in pricing strategy; these include:

- pricing decisions being too biased towards cost structures rather than market considerations;
- prices being set independently of other marketing-mix elements;
- too little account being taken of opportunities to capitalize on differentiation;
- prices not varying greatly between different market segments;
- pricing policy being defensive, i.e. led by events, rather than being one of the prime marketing tools.

## 4.7   Summary

The key message here is about the importance of pricing.

- Price is an inherent part of any product and used by customers to make judgements and decisions.
- Setting price must be considered broadly, not just in context of a calculation of costs.
- Price is a variable and can be used tactically and creatively as an inherent part of the overall marketing approach adopted.
- A systematic approach must be adopted to making pricing decisions and policy, but there is manifestly no magic formula for getting it 'right'.

*Chapter 5*

# Market research and information

Marketing is concerned with meeting customer needs and responding to a variety of circumstances in every aspect of the environment that is the stage for its operations. Some things about all this are known clearly. Others are less obvious and there is indeed a danger of making assumptions about them without establishing the facts. In this chapter we examine how sound information can provide a firm basis for marketing activity, and how the various techniques of market research can provide an objective method of obtaining the pertinent information needed to operate without undue risk.

## 5.1   Marketing and uncertainty

### 5.1.1   *An antidote to risk*

As we have seen, marketing operates in a dynamic arena. The net effect of this can be summed up in one word: risk. Business is an uncertain undertaking. For example, launch a new product (reputedly with a success rate of only one in ten) and the risk is all too obvious. In some businesses, with the long lead times and massive capital undertakings of, say, manufacturing an airliner, this effect is maximized. Management must assess risks in all its decision making and, while there may never be one, provable right answer, what is decided always matters, and there may often be a great deal hanging on the decisions that are made.

Decisions are assisted by knowledge and in this context information really is power; and this brings us to the role of research in business – *market* research. This is a technique – in fact a body of techniques – that provides information to assist decision making in business.

It does not remove judgement and experience from the process. Rather, it supports them by providing a better basis of fact. It can do a whole range of things, from confirming a suspicion to unearthing new, previously un-thought-of, facts. Sometimes this new information changes matters just a little, though maybe significantly. Sometimes it may reverse a situation. Always, the intention is to add to the objective element of decision making, to make it better informed and ultimately more likely to take business in the right direction.

At the end of the day, market research, in all its forms, has one overriding purpose: *to reduce risk and thus increase the likelihood of success.*

## 5.1.2    The purpose of market research

Let us put research in context. Decision making is central to carrying out managerial functions and making the planning and monitoring process work. Good decisions are taken on the basis of availability and use of relevant information. The information of most concern to marketing management comes from markets and customers, present, potential and future, and concerns the shape, size, nature, needs, opportunities and threats within the market. Market research is the means of providing that information.

A full definition of market research, extending beyond that above is:

The systematic problem analysis, model building and fact finding for the purpose of improved decision making and control in the marketing of goods and services.

This implies that research is not just an information tool: it is a means of providing guidance to help improve the abilities of management within an organization, as well as a means of making a contribution to the management of the marketing mix. It can be used to help decide on: the marketing strategy required to meet the challenge of new opportunities; which market gaps to approach; and which are the key areas of interest for future marketing strategies.

The two basic purposes of research are:

- to reduce uncertainty when plans are being made, whether these relate to the marketing operation as a whole or to individual components of the marketing mix such as advertising or sales promotion;
- to monitor performance after the plans have been put into operation; in fact, the monitoring role itself has two specific functions: it helps to control the execution of the company's operational plan and it makes a substantial contribution to long-term strategic planning.

Simply stated, research covers all the 'finding-out' activities of marketing. The methods used may be simple: such as the completion of the questionnaires that you find in a hotel bedroom. Thus, despite what follows, some research can be simply done without vast cost. Collecting and analysing these progressively indicate the views of guests. Or methods may be complicated: data being gathered by post, telephone, email or face-to-face interviews from large numbers of people spread widely, maybe internationally.

Research affects an essential early stage of a marketing process – the identification of consumer needs – and can continue to update those views in different ways as things change over time.

Here, as so often with marketing jargon, we need to be very clear. The description *market research* is applied in two different ways. First, it is an umbrella term for a number of similar, but significantly different, types of research. Second, one of these specific types of research is itself called *market research* – of which more in due course. What sits under the market-research umbrella?

### 5.1.3 Different kinds of research

The term *market research* encompasses five major types of research.

1. **Market research:** This is research that, as an individual technique, investigates markets – asking who buys what, in what quantity.
2. **Product research:** This focuses on the product or service, asking what is right and wrong with the products of the company, or some aspect of them.
3. **Marketing method research:** This examines aspects of marketing activity to see how well it is operating, asking whether communication, distribution, etc. are effective.
4. **Motivational research:** This looks at the way people think, asking about the basic reasons why people buy the products they do and what they feel about them.
5. **Attitude surveys:** These focus on customers' perceptions of, and attitudes to, products and to the companies who make them.

Like any other form of research, market research – in the sense of all the different kinds of research listed above – can investigate only *past* behaviour. Remember, it is not possible to *research* the future.

Research is of course very helpful in *predicting* future behaviour, but research is essentially different from prediction; and this is something that must always be borne in mind. When attempts are made at prediction (e.g. political election opinion polls), serious errors can be made. The fact that research is not infallible, cannot simply unearth exclusively accurate facts, causes some cynicism. One researcher (quoted anonymously at a Market Research Society conference) said, 'Good research reveals things I didn't know. Indifferent research reveals things I already knew. Bad research concentrates on things I know are untrue'.

The role of research, therefore, is to improve the fact basis on which forecasts and decisions are made. It must be made to work hard and to focus accurately on information that does help.

## 5.2 The role of market research

It is worth spelling out in a little more detail the range involved here. Market research provides information that assists an organization to define opportunities for product development and market strategy. It works by assessing whether marketing strategies are accurately targeted, and by identifying market opportunities or changes that are required by customers. Market research tends to confirm issues that are well known in a market initially; but, if planned well and effectively, it will also identify new opportunities, market niches or ways by which to improve sales, marketing and communications activities.

The role of market research, therefore, is to reduce uncertainty in decision making, to monitor the effects of decisions taken and identify the performance of a company or a product in the market.

## 5.2.1   Specific uses of market research

To be more specific, we can list five key uses for market research to:

1.  **identify** the size, shape and nature of a market, so as to understand the market and marketing opportunities;
2.  **investigate** the strengths and weaknesses of competitive products and the level of trade support a company enjoys;
3.  **test out** strategic and product ideas, which help to define the most effective customer led strategies;
4.  **monitor** the effectiveness of strategies;
5.  **help to define** when marketing expenditure, promotions and targeting need to be adjusted or improved.

The variety of purpose listed above makes it clear that market research is not simply a 'first check'. It *is* useful ahead of any action, but it also provides a means of checking and refining views as operations proceed. Companies – especially those for which budgets always seem tight – who have selected one of these uses for market research are always concerned to make the research a worthwhile investment. Best results come when their marketing and sales planning is influenced by the results of research – in other words, when research pays for itself by providing a basis for change and improvement in operational matters.

## 5.2.2   Further possibilities

The range of possibilities for research is considerable. To illustrate further, some of the regular and main reasons for using market research are to:

*   **provide** data on the market, or a market segment, and to discover whether the sector is increasing, staying the same or decreasing in importance to customers;
*   **obtain** information to help to understand who the customers are, and the way in which they buy and use certain products;
*   **evaluate** customer service, assessing what customers feel about the services that they are receiving and their quality;
*   **research** customer attitudes and needs on a continuous basis to discover which product types are selling and where there are opportunities for new sales;
*   **achieve** better targeting, understanding what media and messages influence consumers to buy the products;
*   **identify** changes in the market that will affect how marketing must proceed in future.

These, and more no doubt, give a real insight into the possibilities. Market research can be central to the marketing process. It underpins the activity, grounding it in reality and helping give it the best possible focus.

So, to recap, here are the key things characterizing market research.

- It is a means to an end and can help improve marketing effectiveness and reduce business risk.
- It encompasses a range of different kinds of research and these can be deployed to help in a variety of ways.
- While research employs 'scientific' methods (statistical techniques, among others) it is not infallible. It provides guidance and this supports and enhances the management judgement that is always necessary
- Despite its ultimate fallibility, it is a valuable aid and many aspects of marketing can benefit positively from its help.

## 5.3    The techniques of research

The techniques and methodology that market research deploys are many. Their use demands specialist skills and, like so much in marketing, it is dynamic – not least, the techniques themselves are constantly developing and changing. Their successful use is, in major part, dependent on experience of both market research and the broader issues of marketing. In this section a number of different factors are touched on briefly under a series of main headings that will guide you through the whole complex process. In this way, despite the restrictions of space, we can touch on all the key issues involved in research.

The arrangement is broadly chronological, starting with factors to do with preparation for research, and ending with the presentation of findings and how it can make decision making less uncertain.

The first heading takes us into elements that crop up at the beginning of the process.

### 5.3.1    Planning market research surveys

Lurking behind most research surveys is a problem that needs solving (though this includes 'positive problems' such as seeking to find the best way to take advantage of an opportunity). At the outset of any study it is the researcher's job to determine what the problem is and show how it can be solved. The researcher must develop skills in taking a brief from the 'problem owner' and translating it into a 'proposal' for carrying out the study. In the proposal the researcher states the objectives of the study, the methods that will be used to meet the objectives, the timing, the composition of the research team and the cost. This is true whether an organization is undertaking research itself or retaining a specialist agency.

### 5.3.2    Desk research

There is no point in reinventing the wheel – it is costly and time-consuming. If data exist, they should be used and not collected afresh. Desk research is the collection, sifting and interpretation of published data. It plays a part in most surveys, even if only to use the known breakdown of the population to guide the selection of a quota sample. Elsewhere it may involve the researcher in delving in the library or searching online databases for information on market size and structure. This is an area where the ongoing IT revolution has created many new options.

### 5.3.3   Standards and methods

Standards have had an increasing impact on the practice of market research. In the UK, these relate primarily to the Market Research Society's Code of Practice. Standards do not define good methods, but they encourage quality. They are intended to be implemented at an organizational level, rather than being a matter for individual practitioners – but they need to be followed by practitioners in an organization and are there to promote consumer confidence in the process. After all, without consumers' cooperation, no research is possible.

### 5.3.4   Sampling and statistics

Sampling is a worry for most researchers who are new to the business. The mathematical basis that allows small numbers to be researched with the confidence that they are truly representative of the population as a whole (whatever the group is) is vital to research. Understanding the rudiments of random sampling is necessary, even though in most day-to-day surveys the researcher may learn to trust a quota sample of, say, 300 interviews spread across five cities. Without an appreciation of why and how different samples are selected, the researcher cannot claim to be undertaking a valid, scientific piece of work. Research must therefore be conducted in a way that brings specialized knowledge about this factor to the table.

### 5.3.5   Questionnaire design

Good market research is about asking the right people the right questions. Not much more, and not much less. If we fill in forms, we are soon critical of them if they are the least bit ambiguous, and this is the danger. In theory, questionnaire design should be easy, and yet it is one of the most difficult tasks to get right. Designing questions that draw out accurate information from everyone, that can be completed easily by the interviewee, that flow well and leave respondents feeling that they have contributed something worthwhile should be the aim of all researchers. This is therefore an especially important part of any study.

### 5.3.6   Geodemographics

Time was when survey samples were selected from a representative quota of the population based on sex, social class and age. Over the last twenty years the technique has made it possible to link the characteristics of people with the neighbourhoods in which they live. This has become a powerful tool in allowing researchers to infer certain types of behaviour through knowing the geography of people's homes. Geodemographics gave sampling a new lease of life. This is clearly of most importance to research directed widely across the population as a whole; for an industrial company wanting to research buyers of, say, agricultural machinery of some sort the identification of people to contact is less of a problem.

### 5.3.7 Quantitative research over the Internet

The Internet has become a collection of virtual communities, and all are focused on sharing information. For some organizations the Internet may be no more than a marketing or promotional tool. Others will benefit from operating and distributing their services over the Internet. Market research is a service that now finds the Internet providing a new method of collecting and distributing information. As a result, web and email questionnaires can be useful tools to researchers. The sophistication of this area changes as you watch, as do the applications to which it is put.

## 5.4 Data collection

The next area to consider is the part of the process that actually pulls together the raw data. There are a variety of different ways of going about this; they are not, of course, mutually exclusive and a mix of methods may be used in many research projects. Some methods match very logically with a particular target audience. It may be so numerous that only low-cost methods are possible, or so difficult to contact that special – and more expensive – contact methods are the only way.

### 5.4.1 Quantitative research

Quantitative research is that which supplies a number to anything that can be measured; indeed there is a large body of researchers who argue that measurement can be applied to anything. Quantitative research produces 'hard' data, which can be defended or challenged and are more than just opinion. It is based on sizable surveys, which, in the main, use samples of upwards of 200 people. However, the well-rounded researcher does not see quantitative research as a technique that can stand alone. It is often appropriate to plumb people's opinions, first using qualitative techniques before determining exactly what should be measured. The qualitative and the quantitative methods therefore are often used together in this way, though it is always somehow easier to give findings credence when something measurable is involved.

### 5.4.2 Face-to-face interviewing

The market-research industry has been built around the core technique of face-to-face interviewing – in the street and in the home, and industrial research in the office. It is still the bedrock of many studies, since it allows the interviewer to use personal skills to elicit the information in a way that enhances accuracy. It also allows the showing of visual aids, smoothes the interview and allows deeper insights to be gained than through more mechanistic methods.

### 5.4.3    Telephone interviewing

The telephone rose in popularity as a market-research tool in the 1980s as it allows interviews to be carried out speedily, and under close supervision through central control. This means of contact allows the researcher to sample households easily, anywhere in the country; indeed it is a technique that can allow prompt contact internationally also, albeit at higher cost. It is not necessarily a cheap method; in fact it costs approximately the same as a street interview.

### 5.4.4    Postal surveys

Sending a questionnaire through the post must be one of the simplest scenarios. There is a certain prejudice about the use of postal surveys, a belief that the response they bring is usually inadequate. But this is perhaps only because they are frequently used in the wrong circumstances. They produce excellent results when there is a strong relationship between the respondent and the company carrying out the research. They are suited to testing opinion and sensitive subjects and work best with closed questions. Different styles and approaches are designed to maximize responses.

You will sometimes see a similarity between postal research and direct-mail marketing techniques. Both are susceptible to minor differences for their response. For example, just changing a heading, adding further explanatory text or providing an incentive to respondents makes a difference. Indeed, I once received a very nice pair of sunglasses from a motor manufacturer in the post following my completion of a questionnaire for them some days previously. Such is not uncommon.

### 5.4.5    Omnibus research

Omnibus studies are targeted at certain groups of respondents and are run at regular intervals. They provide the facility for an organization to buy space for a limited number of questions in a large interview programme. Because the cost of the interviewing and analysis is shared among a number of organizations, each contributing questions to the omnibus, it is a particularly cost-efficient means of collecting data. This is a technique that has seen an ever-expanding range of omnibuses covering all manner of target groups spring up over the years.

### 5.4.6    Panels and diaries

A panel differs from an omnibus study in that it is a survey of the same people each time. In practice, it is not always *exactly* the same people all the time, because people drop out and need replacing. Care is taken to replace departing panel members with others with relevant demographic characteristics. The questions that are asked of the panel are consistent so that results can be tracked over time. This provides reliable trend data on purchasing or on such things as television-viewing habits.

Such panels, and they exist all over the world, are usually sponsored by large media or companies wanting to keep a check on movements in their target markets. The panel members keep records of their purchases and activities in diaries; that term is used here therefore in a technical sense.

## 5.4.7 Retail audits

As the name suggests, retail audits take place at the shop or store. By checking the stock turnover at retailers' premises, the audit companies, which instigate this sort of research, produce accurate figures on the market shares of a wide range of consumer goods. The subscribers to the audits can then use the results to monitor changes in brand shares, and within the distribution routes through which their goods are sold, and in turn to adjust their strategies in the marketplace to accommodate the latest situation.

## 5.4.8 EPOS

*EPOS* stands for *electronic point of sale*. It describes a process carried out by scanning the bar codes at the checkout, typically, though not only, at retail outlets. This allows researchers to measure quickly and accurately which goods have been sold and at what prices. EPOS is an extension of the retail audit. It is an excellent means of tracking product data as well as furnishing researchers with a basis for much predictive modelling.

## 5.4.9 Qualitative research

In many studies researchers want to obtain a deep understanding of not just what is happening, but *why* and *how* something is happening. To achieve this the qualitative researcher works with small samples of people, sometimes on a one-to-one basis and sometimes in small groups. These are less like interviews and more in the nature of conversations or discussions. They are long and unstructured and require considerable skill to draw out relevant information, and more to analyse the significant facts from them afterwards. Qualitative research can produce rich data, probing into people's unconscious attitudes and needs. Because the samples are small, there is no attempt to measure responses.

## 5.4.10 In-depth interviews

Using open-ended and unstructured interview guides, the researcher carries out in-depth interviews to 'get beneath' the superficial responses. The in-depth interview permits the researcher to be flexible in the order and style of questioning so that avenues of interest and relevance to a particular respondent can be explored. This can be a valuable technique, but time equals money in all such contexts, so expenditure is increased.

## 5.4.11 Group discussions (focus groups)

In a group discussion between five and nine people are led into an exchange of views by the researcher (who is called the moderator). The interactions between people in the group are used to flush out views that would not otherwise be raised in one-to-one interviewing. The group discussion is a widely employed technique for researching new concepts and guiding creative decision making. Group sessions can yield rich information, but they do require experienced researchers to direct them and make them work and obtain true responses and not just the 'party line'. Here too costs can be high.

### 5.4.12    Hall tests

There are many occasions in market research when it is necessary to have people look at (or touch or taste) a product. For all sorts of reasons it may not be possible for this to take place in consumers' homes. When this is the case, hall tests are set up. Target consumers are 'recruited' from busy streets and invited to a nearby venue (hence *hall*), where the test takes place. A variety of techniques may be used in this context, but all have in common using the product itself as part of the enquiry.

### 5.4.13    Sensory evaluation

Sensory evaluation is a tool to help the technical research-and-development teams design better products. It focuses on a small number of aspects of a product such as the materials that are used in its manufacture, their quality, the shape of the product and its performance in use. Data can be mapped to show where the product stands against consumer preferences and in comparison with the competition. Because the evaluation considers a number of variables, this type of new product research benefits greatly from multivariate analysis.

The profusion of techniques is obvious here, and further study will throw up additional specialist approaches: research into advertising methods, radio listening and more.

## 5.5    Analysis and modelling

Just assembling a 'pile of data', as it were, is of little value in its own right. The data need analysis; they need interpreting to make them able to play a proper part in decision making.

### 5.5.1    Data analysis

Data have no value in isolation. It is the implications of data that really matter – what the data show or suggest. This means that researchers have a responsibility to tease out only the data that are relevant to the objectives of the study, and to simplify them so that the user can quickly and easily see a pattern. The data must be presented in a form that can be understood and, hopefully, they will lead naturally to conclusions and recommendations. This part of the process may be automated and conducted largely by computer, but the job of setting up the required analysis is skilled and important. Only bad research collects large amounts of information and leaves it at that.

## 5.5.2 Modelling

Computers have enabled researchers to get more out of their data than ever before. For example, programs now exist for testing the prices that people will pay for a product. They can show the degree to which consumers will trade off some feature such as quality or design against price. Simulated test markets can be set up. Missing data can be inferred by 'fusing' together sets of data. Data can be analysed to map or segment consumers to show their different characteristics or attitudes to brands. And the models can be used to forecast a course of action. This is an area of considerable complexity, and, as such, it may be beyond the remit of many small organizations, but technological change is making it more accessible all the time.

## 5.5.3 Presentation and reports

It may be an obvious point, but the nature of research makes the presentation of its findings an inherently important part of the whole process and worth a brief comment. The final output of researchers' efforts is the presentation of their work – their findings. Presentations are the 'day of reckoning' for researchers, a chance to make a mark for better or worse. Good presentations have a clear objective. They are short but to the point with little time spent describing the method and more time spent on the findings and conclusions.

The use of visual aids to communicate the data through charts and diagrams has become sophisticated, and the (good) use of PowerPoint means good clear information can be produced quickly and cheaply. However, remember the phrase now seen often: 'death by PowerPoint'. If you want to know how to make PowerPoint presentations in a marketing context, read *Killer Presentations* (Nick Oulton, How To Books).

While the personal presentation has impact, the written report is more enduring and may be read over a long period of time. It must be got right (it may come back to haunt its writer in times to come!). As with presentations, the same rules apply. The audience must be kept in mind and the writing style should quickly and clearly communicate the points, leading logically to the conclusions and recommendations.

This aspect is heightened in importance by the very nature of research findings. Most people cannot grasp complex figures and their implications instantly; and most find a clear graph easier to take a point from than a lengthy computer printout. It is possible, of course, for figures to give the wrong impression and care is always necessary in the presentation of such information. In the case of research, information has been produced at an expense of time and money and it is obviously wasteful if its poor presentation disguises rather than enhances its value.

## 5.6 Knowledge and success

### 5.6.1 Research as a success-generating factor

Let us put everything said so far in context. The risks faced by businesses today have never been greater. Competition is fierce at every level of every market. Small businesses are likely to be funded by a family's life savings or expensive borrowing.

The cost of failure can be very high for the entrepreneur and for their associates. Large businesses face the same risks except that there are more noughts on the figures. The cost of failure for the large business may mean redundancies, scrapped plant and dire financial losses.

Success or failure in business is a consequence of making the right or wrong decisions. The right decisions are easy with hindsight, but much more difficult when the conditions are unknown. It is a relatively simple matter to plan the production resources and estimate the financial requirement for a business. And yet both these plans must be based on understanding the needs of the market, and on whether customers will buy the products and then become repeat buyers.

It is those market needs that are most often misjudged, assumed or even taken for granted. Uncertainty about what the market wants, both now and in the future, is one of the most difficult problems with which businesses must cope. More than ever, decisions in business require robust information. If information on markets is a key to business success, it follows that the people who can supply it hold considerable power.

It is the role of market researchers to provide sound information to guide business decisions, set strategies and monitor the implementations to give feedback or say whether it has been successful or unsuccessful. And, one might add, it is the role of marketing people and senior management to demand or organize such information. The techniques available to researchers have been developed and polished, especially over the last three decades. There is no area where market-research techniques cannot be used. They are as useful in social marketing to probe why people drink and drive as they are to manufacturers selling alcoholic drinks; as useful to the government trying to obtain recruits for the armed forces as for theatre managers trying to measure their audiences' likes and dislikes. The skill of the market researcher is not just being able to apply a special technique, but also knowing which to apply and when.

## 5.6.2  A research project – an overview

There is, as we have seen, a good deal of complexity about research, perhaps particularly in the area of methodology. However, the overall structure of a research project does follow a pattern and a snapshot of this will help put the process in context. The following encapsulates the process and illustrates its use in practice. It describes a project conducted by a market-research agency (though things would not be changed so much by its being an internal project).

Six key stages are involved.

### 5.6.2.1  Objective setting

This is a necessity if the research is to have a clear focus and not to get out of hand, either trying to look at everything or looking selectively and not at what will produce the information required. Objectives should always be: specific, measurable, achievable, realistic, and timed. It is as important to be clear on boundaries and what will *not* be done and on what is to be done.

With objectives clearly in mind, the next stage follows logically.

### 5.6.2.2 The brief

It is sensible to have a statement in writing that summarizes the project. This prevents misunderstandings, especially when a number of people are involved. The brief should state clearly:

- the objective, commented on above
- the problem the research should address, or indeed the opportunity
- the business context in which the project will take place
- something about the method to be adopted

### 5.6.2.3 Designing a research programme

With an agency-conducted project, this will be described and summarized in the research firm's proposal and normally follows discussions with them about the brief.

The possibilities and limitations of the programme must be clear and a typical proposal will set out:

- **The background:** This refers to the reasons for doing the research, specific objectives and the areas of information to be investigated (and indeed not included);
- **The methods**, or often the mix of methods, to be used: These may vary from telephone interviews to personal interviews. Here too there should be a clear description, if necessary, of the categories of people to be approached. Note that there is no one 'right' mix of research methods for a project. It is possible – probable – that different firms pitching for work will suggest a different methodology mix and suggestions need weighing carefully before a final decision is made;
- **The timetable:** As with any project, the stages and timings need to be set out;
- **The costs:** These are the fees and expenses that the agency will charge for the prescribed work. More than one quotation, as has been said, may sensibly be sought. Research can suffer from the wrong financial view: too high a cost may just be a firm overengineering to inflate the profitability from a potential project, but too little being spent may render the scale of the project too small to be able to produce meaningful results.

### 5.6.2.4 Fieldwork

Then the right questions must be asked of the right people. An appropriate statistical sample may be needed, and this will need to link appropriately to the methods to be used. The work itself (perhaps followed by some desk research and the capture of existing data) can take place and the information it produces can be recorded.

### 5.6.2.5 Data analysis

This is the 'number-crunching' element of research, most often involving computer analysis. The captured data need to be sorted, for example tables produced showing the total response and that of individual groups within the total. Precision is necessary here, and at the end an accurate picture must be able to be set out.

### 5.6.2.6    Reporting

Any research project of any size is likely to result in a written report. In addition, this is usually summarized by an agency in a formal presentation. The report of the research must:

- present the findings clearly, concisely and in a way that links specifically to the original brief;
- draw conclusions and suggest action, or at least begin to do so (some agencies are hardly involved in this stage, just presenting the facts – indeed that may be what a client wants; others make a virtue of seeing the analysis through).

Whoever does what at this stage, it is the final analysis and consideration of conclusions and an action-oriented conclusion to the project that give it its real value. Whether the action is low-key or high-profile, it is this that makes research worthwhile.

While projects vary in nature and scale, this broad approach would be descriptive of many and certainly gives an idea of the process involved.

## 5.7    Summary

Market research should not be seen as a panacea able to cure all business ills, but as a valuable asset in the battle for an organization to survive and prosper in the harsh commercial world.

- It is a complex, specialist technique that demands that the right mix of methods be used in the right way (something that may also demand specialist expertise).
- It can assist – or rather its findings can assist – in the decision-making process, but does not replace judgement, which is also necessary alongside the interpretation of research findings.
- It is only ever as good as its brief: ask the wrong questions of the wrong people and it should be no surprise if the findings are neither reliable nor useful.

Lack of research, or bad research, can seriously handicap – and at worst cripple – an organization. For many businesses, it is not just a desirable option: it is an operational necessity. Without it they are effectively operating with one arm tied behind their back.

*Chapter 6*

# Routes to market: distribution channels and methodology

Marketing links the organization and the world outside it. There is a huge gulf between a production process and a factory, and customers actually buying, using and finding a product satisfactory. Distribution is the umbrella term for everything that links an organization with the outside world. In this chapter we review the processes and techniques that set up a route (or routes) to market; making the right decisions in this area and making the chosen methodology work well are integral to effective marketing.

## 6.1 The way to market

### *6.1.1 The need for distribution*

Marketing must link to the market, not just in terms of having a focus on customers and their needs, but literally. This link, involving the *place* P of marketing (see Chapter 1, 1.3.1, where we discussed the various Ps), is provided by distribution.

Distribution channels are the more specific concept with which marketing people must get to grips. This is the network of organizations necessary to distribute goods or services from the manufacturers to the consumers. The channel therefore consists of manufacturers, distributors, wholesalers and retailers. Marketing people also talk about *chains of distribution*, which also describes this area and implies the various kinds of intermediary involved.

However good any product or service, however well promoted and however much customers – and potential customers – want it, it has to be put in a position that gives people easy access to it; it must be distributed. And this can be a complex business, though it is certainly a marketing variable, albeit one that many regard as rather more fixed than it is. Consider the variety of ways in which goods are made available. Consumer products are sold in shops – retailers – but such may vary enormously in nature, from supermarkets and department stores to specialist retailers, general stores and more – even market traders. These may, in turn, be variously located: in a town or city centre, in an out-of-town shopping area, in a multistorey shopping centre or on a neighbourhood corner site.

But the complexity does not stop there. Retailers may be supplied by a network of wholesalers or distributors, or they may be simply not involved; some consumer products are sold by mail order, or door to door, or through home parties (like Tupperware). A similar situation applies to services. Even traditional banking services are made available in stores, from machines in the street and through post and telephone – even on a drive-in basis. Banking is also an example of the changes going on in the area of distribution, many of these developments being comparatively recent moves away from solely traditional branch operations. Business-to-business, industrial products are similarly complex in the range of distributive options they use; and overseas markets add complexity.

New channels of distribution now exist as a result of the information-technology revolution and the advent of various forms of e-commerce.

Similar chains of distribution exist in the marketing of every kind of product. The terminology may be a little different, *distributors* rather than *retailers* in industrial marketing, but the principles are essentially the same. The section on market maps, which follows shortly, will explain further through an example.

Any such example is simplistic. The various possible types of entity involved and the chains down which things flow can vary considerably, as can the complexities involved. Hence *channel management*: this is the process of managing the distribution of products and services and the various external organizations and people who may be involved.

## 6.1.2    Channel management

Distribution is crucial to marketing success: products must be accessible to customers and, however well advertised they are, this is a prerequisite of their being purchased at least in any quantity.

But distribution is more than just setting up a mechanism to make goods available. It also acts as part of the process that persuades people to buy. In other words, the quality of distribution – how well the various channels work – is a marketing variable. Managing the distribution process is a creative marketing activity, which is conceptually no different from implementing a creative advertising campaign. There are several separate, but linked, aspects to channel management; let us summarize some of them.

### 6.1.2.1    Channel choice

Decisions must be made about which channels will be used (and which will not). Note that channels are not mutually exclusive: several may be utilized together and the mix (and relative importance of different channels) is what must be fixed on. In addition, this is not choice as it were from a shopping list. Some channels are common, and utilized in a standard form. More often management is involved in creating a unique interpretation of what is done – their version of the utilization of that channel – and, sometimes, in creating new approaches entirely.

### 6.1.2.2  Directing marketing activity towards different channels

Different channels involve accessing customers in different ways and may need different approaches and techniques. The sales element is often important here. For example, one channel may contain customers who can be handled successfully by the field force; another may demand a more sophisticated approach using key account managers.

### 6.1.2.3  Managing the channel at unit level

If the channel consists in part of individual elements – for example some sort of distributor (organizations buying in something for subsequent resale, say) – then an individual relationship needs to be created between the parties. The organization must prompt them to do the best possible job on its behalf. This may involve everything from the provision of information to motivation and incentives for those involved.

How a company analyses the distributive possibilities and organizes to utilize a chosen method or methods effectively is certainly important to overall marketing success. It is affected too by any overall trend in the marketplace. For example, if out-of-town shopping centres are thriving, it may make sense to use them; or not.

Success is also in the details, everything from the strategic view taken of channel management to the tactics of working through a particular individual distributor perhaps. Because the elements conditioning this area of marketing are complex and interlocking, one thing affects another and, at the end of the day, it is the overall mix that matters and how it is managed. The key issues are to:

- make good **channel choice decisions** and never regard them as fixed for ever;
- create **strategic marketing approaches** that make sense for each channel;
- manage the **relationships** involved, especially where success is dependent on other people along the chain over whom no direct control is possible;
- throughout the process, **focus on the customer** – no scheme and no channel will bring commercial success unless it is satisfactory in consumer terms; and ideally, of course, customers must find that it meets their needs and is entirely to their tastes.

### 6.1.3  The impact of distribution on marketing success

To see how important distribution is to marketing – positively or negatively – one has only to consider how it evidences itself in practice. What is described in Panel 6.1 is probably in essence a common occurrence.

### Panel 6.1  A negative distribution experience

Consider a personal example: I am considering adding a scanner to the computer equipment in my small office. My main requirement is that, as well as pictures,

> it will scan text and allow it to be converted into a Word document and be edited. I visit several specialist outlets. All have a range of machines on show. Will they do this? No one knows! There is no offer to find out or to try it and see (if they could demonstrate it, I would buy one on the spot). I retreat back to the office and telephone a large computer firm, the manufacturer of some of the machines that I have been looking at. Told of what occurred they express no surprise whatsoever. They do offer a demonstration on their own premises, but I am still trying to find the time to attend.

The example in Panel 6.1 is that of a distributive channel not working well, and perhaps of one being badly managed, too. Distribution must be got right, and this means working at it.

### 6.1.4   *The availability options*

Products may be available very widely. Consider the simple, everyday example of a chocolate bar. Outlets where this may be bought include: grocery and provision stores of all sorts; all types of supermarket large and small; confectioners; newsagents; petrol stations; cinemas and other entertainment centres (e.g. a bowling rink); leisure centres and clubs; some pubs and bars; vending machines; and cafés and sandwich shops.

To this one could add a list of different locations for some of these that would include the high street, railway and bus stations and out-of-town shopping centres. This is very different from an example such as Wedgwood china, which is sold through exclusive dealers and might be available in only one shop in a particular town. Yet both approaches are valid marketing strategies.

Areas where change has occurred, or is occurring, can prompt rapid customer reaction if whatever new means are offered are found to be convenient. Old habits may die hard, but if change is made attractive and visible then new practices can be established and these too can then quickly become the norm – as, for instance, with the rapid uptake of Internet shopping in some sectors. Conversely, if a distribution method is inconvenient for customers, they will seek other ways of access to the product (or simply not buy it). However, sometimes inconveniences are tolerated because need is high, or is compensated for by other factors. For example, there are some who will put up with queues in one shop only because the alternative is too far away or has nowhere convenient to park nearby.

### 6.1.5   *Market maps*

An organization can use the concept of *market maps* in various ways to help both in the process of channel selection and in managing disparate channels. What's this? It is a graphic representation of the routes a product travels along, which can be used to simplify and make more precise the job of channel selection and management, particularly comparison between different channels.

This concept of market maps – originated by Michael T Wilson in his book *The Management of Marketing* (Gower Publishing) – formalizes simple flow charts into

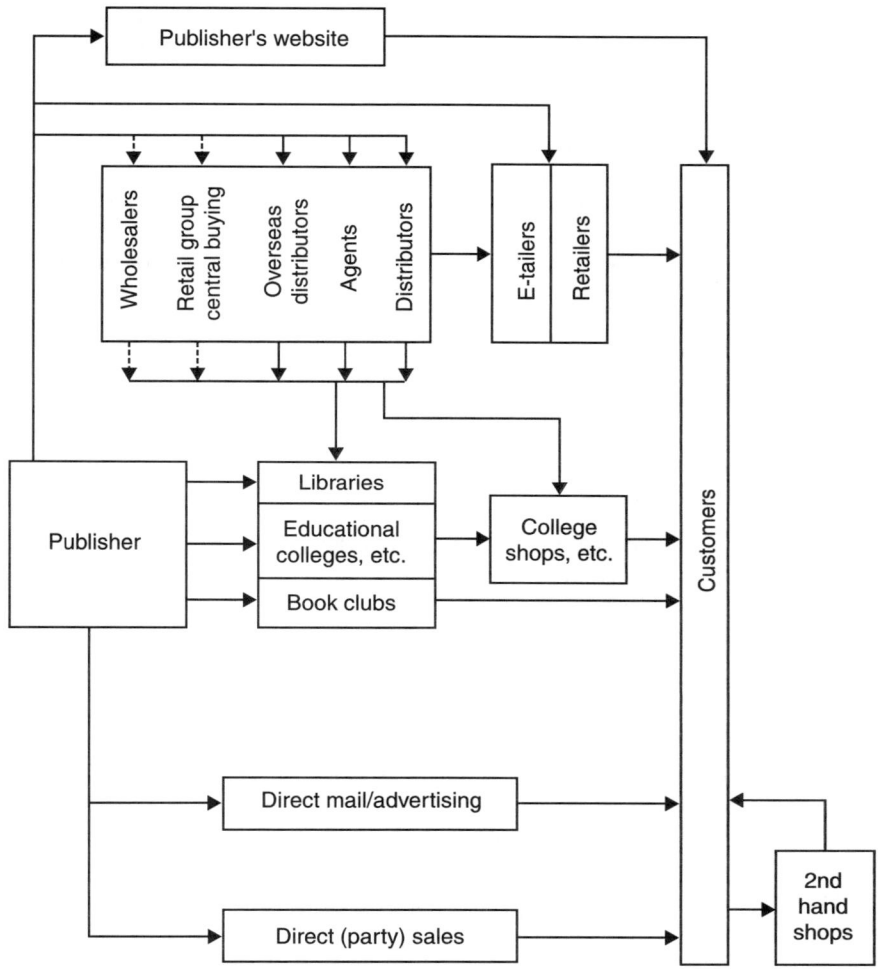

*Figure 6.1 Market map for book publishing*

specific form. It aids analysis, planning and implementation. It describes the nature of the system, shows all the channels in existence or use and can be quantified to show what is happening where.

A simple example of a map, showing the various channels involved in the book publishing industry is shown in Figure 6.1; where you obtained this book should be on there somewhere.

Preparing a map means:

- listing the **categories** of consumer or end-user, and any subdivisions they may contain;
- listing any additional **influencers** (e.g. people having the role that architects do in specifying building materials);

- asking **questions** about customers and their characteristics so that information is clear alongside how their purchasing relates to the map, for example:
  - Who are they (male/female, age, buying power, etc. for individuals and comparable information for industrial buyers – e.g. type of industry)?
  - What are their needs (e.g. for value, performance, convenience – such factors can be linked specifically to a particular product)?
  - How are their needs being satisfied (by both direct and, if relevant, indirect competition)?
  - Where do they buy (linking to the different channels featured on the map)?

This process of preparing a map may necessitate assembling significant amounts of data and ensuring that sales figures can be produced accurately in the right form; even some research may be useful. So be it. The detail is worth assembling and, with the right setup – for example computer programs that will link conventional sales data into 'channel form' – much of it can be regularly updated very easily. Given the information, and a map to illustrate what is happening, there are several benefits stemming from the approach.

- It assists planning and setting strategy (which channels to use/not use).
- It monitors performance, allowing action to be taken to fine-tune marketing action directed at specific channels.
- It highlights the relationships involved (which customers use which channels etc.).
- It allows a view to be taken of matters such as pricing and profitability that reflects what is happening, not in an overall sense, but in the way individual channels work.

It is an instrument on which various tunes can be played, and can assist in a variety of ways. The following example shows just one aspect of this, and illustrates how reviewing channels can lead to fundamental changes in marketing approach.

### 6.1.5.1   How channels affect marketing approaches

Returning to the book-market map, shown in Figure 6.1, consider just two aspects of this.

First, consider books sold to traditional retailers, where the criteria for buying are well known. They vary by category of book, of course. Textbooks, say, may be judged in terms of their price, up-to-date nature, link with particular courses (a book officially recommended by a university is more likely to be bought and stocked than one just rated as 'useful'), accessibility, etc., and by the reputation, standing and presentation of the publisher (and author) and more. Buyers will tend to have common views and what information needs to be put over to them can be addressed by the field sales team without (great) problem.

Second, consider universities. Their bookshops may have similar criteria. But their academics have a different perspective and need handling differently. Yet they can be approached to try to secure adoption – agreement that a book become recommended reading. Opening a channel to this category of recommenders may increase sales through two channels: university and college bookshops and larger bookshops with academic departments. The link with such people is, however, radically different from that with the typical bookshop buyer, albeit those who buy academic titles. It almost certainly needs a different sales approach; it may even need salespeople with different characteristics. But it is a channel that can be identified; an approach can be tailored for it and the information coming from updated market-map views can assist in identifying this sort of situation, and in monitoring progress.

It must always be borne in mind that it is a dynamic situation that is being viewed in this way, hence the necessary resolve to review on an ongoing basis. Academics may respond to presentations and want briefings of a particular sort, but they are busy people – how will they use the Internet in their task of finding and assessing suitable texts for the courses they teach? And what response will that need? Real consideration is necessary here.

## 6.2   The case for distributors

There are a number of good reasons for delegating what is an essential element of the marketing mix and using distributors. Let us consider some examples.

- Distributive intermediaries provide a ready-made network of contacts that would otherwise take years to establish from scratch at what might be a prohibitive cost. Clearly, even a large company might balk at the thought of setting up their own chain of specialist shops, and the incidence of this is very low.
- Distributors are objective and are often not tied to one product. They can offer a range that appeals to their customers, electing to pitch this wide or narrow (and there some shops that sell very narrow ranges, e.g. only ties, books on sport, coffee and literally nothing else).
- Distributors provide an environment that the customer needs in order to make a choice. If different competing brands need to be compared, then the customer in an outlet offering a range of similar products can conveniently do this. If a distributor stocks a product and is also well known as an organization with an attractive image, this may enhance the overall attractiveness of the shopping experience in the eyes of a consumer by association. In many fields, allowing potential customers to view a wide choice – as in, say, selecting a television set – is an important aspect of encouraging sales.
- Distributors can spread the costs of stocking and selling one product over all the items they carry, thereby distributing it at a lower cost than a supplier operating alone.
- The cost of bad debts is sometimes lower than it would be otherwise, because the distributor effectively shares the risk (however it may seem sometimes! – slow payment seems endemic in so many industries).
- Since the distributor is rewarded by a discount off the selling price, no capital is tied up by the principal in holding local stocks, though overlong credit can (usually does?) dilute this effect.

- Distributors can have good specialist knowledge of retailing or distribution, which the principal may not possess; or they should have – this clearly varies across different kinds of intermediary.

So far so good, but there is another side to the coin. There can be conflicts of interest between principals and distributors.

## 6.2.1   Distributive downsides

Potential problems may include distributors ...

- ... *who are not as committed to a particular product as its producer is.* If the customer prefers another, they will substitute it. For example, if a customer asks for advice in a travel agency, saying, 'I want to arrange a weekend break and I see there are some good deals in France,' they are just as likely to end up finding themselves going to the Channel Isles. As the conversation progressed, the agent had no particular axe to grind in selling one destination rather than another – and in some cases discount structures may lead them to make particular recommendations that benefit them. In this case perhaps the company whose advertising successfully brought the customer into the shop in the first place loses out.
- ... *who may use the manufacturer's product for their own promotional purposes, something that is often linked to price-cutting.* Not every product whose price is cut wants it (a manufacturer may feel it dilutes image, but be powerless to stop it being done – distribution is replete with legal considerations beyond our brief here).
- ... *who may drop the product from their list if they believe they can make a better profit with another line.* This will clearly affect directly competing lines.
- ... *many of whom expect the manufacturer to stimulate demand for the product.* An example might be advertising or providing display material. Sometimes they are more interested in the support than the product itself.
- ... *many of whom are tough on terms and have complex, time-consuming, ordering procedures.* At the same time they distance themselves from collaboration that could perhaps increase sales for both parties.

The question of whether to deal directly with the consumer is, therefore, dependent, first, on the availability of suitable channels and the willingness of buyers within them to add additional products to the range they sell, and, second, on balancing the economies of the distributors' lower selling and servicing costs with the disadvantages of not being present at the point where customers are making their decisions, and thus having less control over the selling process. Realistically, many companies have no option but to go through existing channels (whether these involve shops or not), though exactly how this is done and the mix involved can be varied. In addition, more radical variants may need to be found and run alongside (and without alienating) the retail chain; this can certainly be a way of increasing business.

## 6.3   Selecting the right channel of distribution

### 6.3.1   Analysing the options

There is no one easy and obvious route for most suppliers. A mix of methods needs setting up, and any decision must be based on facts and analysis. Although this is, as we have seen already, a decision involving some complex, interlocking issues, six main factors will influence the route taken.

1.  **Customer characteristics:** Distributors are generally required when customers are widely dispersed, there are a large number of them and they buy frequently in small amounts. This is certainly true of many sectors of everyday products, less so or not at all with regard to some specialist items.
2.  **Product characteristics:** Direct distribution is required when bulky or heavy products are involved. Bulky products need channel arrangements that mini-mize the shipping distance and the number of times they need to be handled; even a brief look at physical distribution costs shows the importance of this factor.
    Where high unit value can cover higher unit selling costs, then any manufacturer can keep control over distribution by dealing direct, as with certain off-the-page or catalogue-distribution systems, or, at the far end of the spectrum, the likes of those who sell door to door. Finally, products requiring installation or maintenance are generally sold through a limited network, such as sole agents.
3.  **Distributor characteristics:** Distributors are more useful when their skills of low-cost contact, service and storage are more important than their lack of com-mitment to one product or brand. If very specific support is necessary, then other options may be preferred.
4.  **Competitive characteristics:** The channels chosen may often be influenced by the channels competitors use, and there may be dangers in moving away too far and too fast from what a market expects and likes. The competitive interaction in this way between retailers is another variable. In the area of fast food, Burger King tries to obtain sites near to McDonald's; on the other hand, some manufacturers, such as Avon Cosmetics, choose not to compete for scarce positions in retail stores and have established profitable door-to-door, direct-selling operations instead. Similarly, major chains may seek to open branches near existing smaller, independent retailers, not only to take advantage of their market knowledge – they are in an area where there is a demand – but with the aim of replacing them all together. This last may well not be entirely in customers' interests and illustrates one aspect of the sheer power of major retailing groups.
5.  **Company characteristics:** The size of a company often correlates with its market share. The bigger its market share, the easier it is to find distributors willing to handle the product – thus, even a small shop is likely to find a corner for major brands, but will be more selective about what else it stocks. It may not be able to stock everything, but will find space for anything it believes in. Where there is clearly profit to be made, no one wants to miss out on it. Similarly, a supplier

may be innovative (and/or build on a strength) and seek ways of becoming less dependent on the normal chain of distribution.

Creativity may have a role to play here. For instance, cosmetics may sell well in outlets that simply display them, but stores putting on makeup demonstrations (or letting the manufacturer do so) may create an edge – for a while.

Additionally, a policy of fast delivery is less compatible with a large number of stages in the channel and there is a danger that slow delivery (measured in market terms) dilutes marketing effectiveness. As service standards increase so there is less room for anyone who lags behind to do as well as they might – slow delivery is increasingly not tolerated. This has been a problem for some in e-tailing. The instant nature of the purchasing process seems to prompt feelings in consumers that delivery should be swift – a feeling encouraged by those who do just that. Distance (and, for some people, the still new method) can reduce credibility and trust. One bad experience of delay is enough to rule out any repeat business (especially if checking up on any pending matters is difficult too).

6.  **Environmental characteristics:** Changes in the economic and legal environment can also bring about changes in distributive structures. For example, when the market is depressed, manufacturers want to move their goods to market in the most economic way. They thus cut out intermediaries or unessential services to compete on price and deal direct. Again, legal restrictions have been introduced in the UK in recent years to prevent any channel characteristics that may weaken competition.

Also influencing how things are done are overall trends within retailing. Out-of-town shopping, the use of the car (or restrictions on it) and everything from the cost of renting retail accommodation to the desirability of an area influences the likelihood of shoppers patronizing a particular area, and thus a particular shop. Certainly this may influence where all sorts of product are bought, and this in turn may influence what is bought.

An example might be the development of Covent Garden vegetable market in London. Now an attractive area of restaurants and entertainment as well as shops, it attracts people from far and wide. Someone might well buy something, a present, perhaps, in a shop there, choosing something different from what they might have bought if they had shopped somewhere else. Such decisions are based on what is there, how it is displayed and more.

A major retailing trend is towards out-of-town shopping centres of various sorts. In some, small, independent shops fit in well. In others the big retail groups predominate and the environment is not right for the small shopkeeper.

The bigger the environmental change, the more likely it is to have repercussions, and there are doubtless plenty of changes still to come in this area.

### 6.3.2  *Variety and change*

Usually it is possible to identify several different types of potential channel or distributor. In certain industries some of the alternatives may be further from standard practice than others, but that does not mean they are not worthy of consideration, or cannot be part of the distribution mix. Things that are normal now may have originally been difficult to establish. For example, party selling is well known (who does not know Tupperware?) but some companies use this technique in areas where it is very much not the norm and make it successful, as Usbourne Publishing does with children's books.

Some companies are, of course, bound to the standard form in their field, but the point here is that it pays to remain open-minded. Channels may change little, traditional routes may remain the most important, creating the greatest volume of business, but other possibilities may still create some growth.

There are still without doubt many new possible innovations in prospect for distribution (the Internet, to name but one of current interest), and things that seem unlikely today will no doubt be looked back on in years to come as entirely normal. We all have 20/20 hindsight. The trick for suppliers is to make sure some marketing time, effort and thinking goes into exploring and testing new methods. This is, of course, true of most things, but distribution is a prime candidate for the very reason that many do see it as essentially static, at least in the short term. Perhaps this just means there is all the more possibility of using it to steal an edge on more conservative competitors.

So, alternatives need be explored to see which channel or combination of channels best meets the firm's objectives and constraints. However, the best choice of channel must take into account the degree to which the company can control, or at least influence, the distribution channel created. But first, it is important that any channel works from the point of view of *customers*.

### 6.3.3   Service to customers

Brands exist to create superior customer value and thus maximize profit opportunities. Channels have a similar role: they should provide not just a route along which products and services are made available, but should actively create availability in a way that provides customers with a satisfactory – indeed, for consumer goods, interesting, perhaps even exciting – way of obtaining what they want.

Channels must be judged on the basis of how well they deliver their 'package' of service benefits. To define this think about what a channel provides under the following headings.

- **Convenience:** such factors as how much time is taken up, so think about being nearby, having easy parking and sensible opening hours, etc.
- **Range:** choice and mix of what is available, allied products, etc.
- **Price:** value in price terms, includes all aspects of price (e.g. discounts, payment terms).
- **Quality:** overall value, including price and service.
- **Service:** customer-care attitudes and practice, and advice if necessary (includes indirect service factors such as the provision of lavatories in a shopping mall).

- **Environment:** everything from style to cleanliness, also ease of use (e.g. pushchair friendly?) and level of crowds.
- **Identification:** clarity of purpose, e.g. is a shopping centre all high-price or the reverse? For instance, saying 'factory outlets' clearly identifies what is in store at least in a general sense.
- **Image:** in the overall sense of *projecting* such factors as quality and the kind of person to whom it is directed.

Note that channels are subject to segmentation just as much as products. A channel may well focus on a part, sometimes a tightly focused part, of a total market. Thus, in terms of both price and people, channels may be designed to attract in different ways, so you can, for example, buy your new kitchen packed flat in boxes from a chain store at an out-of-town shopping centre, or have someone come to your home and advise and supply you individually, before installing it for you.

Additionally, the channel chosen can directly be used to enhance image, for example:

- upmarket theme pubs featuring more expensive 'designer' beers and speciality drinks use the quality or trendy image to tie in with the appeal of individual products;
- the restricted availability of a top brand of perfume or fine china, sold only through a small number of exclusive dealerships, enhances their image of desirability;
- even a humble loaf sold in a small 'real' bakery has sufficient added to its appeal to command a premium price.

At the end of the day what works for consumers is the ultimate measure. A channel will not continue to operate successfully without the approval of the customers who patronize it (unless they have no alternative, in which case it is vulnerable to competitive forces).


## 6.4  Powerful distributive forces

### 6.4.1  The nature of distributor power

When few companies dominate an area of distribution, they can wield considerable power. In the USA and the UK, among other markets, this is certainly dramatically the case in food marketing. If you sell a branded food product in the UK, most of your retail market will go through just five organizations whose chains of shops make up some 80 per cent of the market. Miss out on one and a very significant amount of the potential market is lost.

This gives such organizations considerable power (see Panel 6.2), and delicate negotiations may be necessary to create a balance that gives a sound basis for doing business to both parties. In such a situation margins are constantly under pressure and yet there is a need to support the buyers in a way that enhances the business opportunities inherent in the chain concerned.

**Panel 6.2 Distributor power**

The kinds of thing any manufacturer might be pressed on include:

- additional time from the field sales force (for instance to help merchandising);
- discounts (and there may be many different bases for them, e.g. quantity bought or when purchase is made; and some are retrospective);
- any special packaging and packing;
- delivery (maybe to multiple locations); labelling; credit terms (and beyond);
- returns and damage arrangements;
- advertising and promotional support;
- merchandising materials;
- training of customers' staff;
- financing (including special credit terms).

These sorts of cost are, of course, all in addition to normal production and distribution costs. Yet major players can make demands here that quickly put margins under pressure, knowing that the pressure for the supplier to maintain a relationship with them is intense. On the other side, a buyer – say a retailer – does not want to alienate a supplier and miss the opportunity of profiting from selling a good product. So, realistically, a balance is necessary and the two separate interests must be made to work; it is, however, one that the supplier may sometimes think is tending to be one-sided.

In some fields, e.g. foods, many suppliers may feel that they are too much at the beck and call of the retailers, especially the large ones. Large customers need careful handling: they are not just different in size, they are different in nature. This links to the area of key account management, which is picked up in a later chapter.

Thus the first stage of channel management is to select the right channel or, more likely, to select and balance the right mix of channels. The second is to make your use of it as effective as possible.

### 6.4.2   An agreed deal

Many arrangements with distributors are contractual. This is not a legal nicety: it is imperative to have the detail right and, in international operations, have it on a basis that is right market by market. This is a technical area where the cost of specialist advice is well worthwhile.

In one European industrial-component company, trading across the world through a long list of national distributors operating in their specialist industrial field, a review of the contractual arrangements across the operation was conducted. Arrangements had grown up over time. Markets had been gradually added and in each case a deal had been struck, by a range of different people at different times, to secure the services of what was regarded as the best local dealer with which to work.

As a result, not counting minor differences, 18 different forms of contract were found to be in force. Early ones in some cases still contained clauses and terms that were not easily compatible with current operations. Worse, distributors were themselves raising issues – particularly of the unfairness of their arrangement compared with another – and this was leading to time-consuming and costly renegotiating. The whole thing needed to be sorted into some sort of order. Better to set a consistent policy in the first place, and have a regular review process in place from the beginning.

## 6.5    Managing channels

### 6.5.1    Active management

Not only are chosen distributors likely to work better on behalf of any manufacturer if communications, support (such as information, training and services) and motivation are good, but they will have their own ideas, and a good working relationship must be adopted if both are to profit from the partnership. At best, all this takes time, and often it is easy simply to see people as sources of revenue, rather than someone to work with. Yet the best may be got from a market only when the two parties do work, and work effectively, together.

For example, among retailers, few will even consider taking on a new product (or taking on board any idea) unless they can be convinced that the demand exists, and that its doing so is more that an optimistic gleam in a supplier's eye. They need to know which market segment the product is aimed at and whether it fits with their customer franchise. And exactly the same is true of industrial and other markets.

Working through any channel of distribution demands:

- clear **policy** – that all parties have clear, understood and agreed expectations of each other;
- clear **terms** of trade (discounts and all financial arrangements);
- sufficient **time and resources** to be put into the ongoing process of managing, communicating with and motivating those organizations and people upon whom sales are ultimately dependent.

This last area of communications, motivation and support, is most important. It is not something that can be approached ad hoc. It needs a systematic and ongoing approach, one well tailored to the people on the receiving end of the communication, especially if something like widespread global networks adds to the complexity.

Selecting and managing channels of distribution demand analysis, decision and activity at every level:

- analysing the market as a whole, how it operates and how it can, and might, be accessed;
- analysing individual distribution channels to see how they fit with your activity and your marketing strategy;
- similarly analysing how well they meet customer needs;
- selecting the right mix;

- prioritizing them in terms of the flow of business through each and the respective weight of activity that you will put behind each one;
- making suitable arrangements (and, if necessary, contracts) with the parties involved (wholesalers, distributors – whatever);
- creating and maintaining a relationship with them through a suitable communications programme;
- planning and implementing individual key account strategies with all those individual customers who warrant attention at this level;
- reviewing performance and fine-tuning activity as necessary;
- thinking and acting creatively to find and explore new channels and new ways of making existing channels work more effectively in future.

It is at the interface, whatever form that takes, that sales are ultimately made. So the supplier must work effectively with what is necessarily a given (and this includes dealing with aspects of the business that you would rather were different), and also seek new approaches and methods where appropriate and take an innovative and creative approach to the whole process of distribution.

## 6.6 Key approaches to making distribution work

As we have seen, channel management consists essentially of taking a marketing approach to the area of distribution. Decisions about *what* to do are important, but so too is what is done to make distribution work and that too defines managing channels.

The following highlights what is necessary and identifies the core activities.

### 6.6.1 Making the selection

The first principle to recognize here is that there is a selection to make. Just because things are performing well, it does not mean that existing channels are right, much less that they will be right for ever. There may be some routes you can rule out at once for obvious reasons; though even 'obvious reasons' may sensibly be questioned and other routes all need consideration.

There are numbers of factors to balance. These include customers, and different customer types; product characteristics; distributor characteristics, i.e. what skills and resources they 'bring to the party'; competition, and the question of copying, matching or avoiding what it does; your own organization, e.g. what is it better to do yourself? what is better to leave to a distributor? And overall economic and environmental factors affecting your product and markets must also be allowed for.

Research and analysis may be necessary here and some options that may not seem so far apart (e.g. exclusive or nonexclusive dealers) may in fact have important differences that demand investigation to enable the right choice to be made.

Remember too that many options have legal consequences. If you sign an agreement with a particular distributor, say, then you may have to live with it, at least for a while. You do not want to fall foul of a problem that simple checking could have prevented.

So, the key things here are to:

- review the options widely;
- gather the necessary information;
- take time to study the options and the information you have about them.

The whole process may need to be repeated literally market by market if operations are international.

### 6.6.2    *Deciding the channel mix*

The point has been made that channel choice is not a matter of picking *the* route to market; that is right for few organizations. It is a matter of picking a *mix* of channels that make sense and deciding the relative importance of each.

This is not simply a matter of ranking them. It means making decisions about your investment in each and forecasting what each will do for you. Here the market-map concept helps clarify what may be a complex picture. It allows the total picture to be taken in, and the way one channel performs to be reviewed alongside others. Channel mix must reflect a clear strategy that is a clear strategy for *each* channel. As different channels may be used for widely different purposes, *every one* must be managed to achieve its own particular objectives.

As an example, a manufacturer of motorcar tyres can sell through two completely different channels. One aims to sell tyres to the manufacturers of cars, which need tyres as original equipment. The other aims to sell car owners replacement tyres. The two are clearly different and also clearly need different approaches, one to access the comparatively small number of motor manufacturers, the other the numerous and different kinds of outlet selling replacement tyres (from garages to specialist tyre centres).

Deciding the mix means making decisions about which channels to use for what (and which not to use), about how to maximize the effectiveness of each channel and about priorities between the number of different routes involved.

### 6.6.3    *Ensuring a customer focus*

Choice of channels is not just a matter of convenience or of finance (or indeed of convention). Channels are there to facilitate the process of purchase. If customers do not like something, then, especially in an age of choice, they will vote with their feet. And this principle is certainly true of channels. Distribution links to the four Ps of marketing; the fourth is usually defined as place. But there is more than just location to worry about here.

- **Location:** This is, of course, important: does the customer want to buy at home, in the high street, at an out-of-town shopping centre, or somewhere else (e.g. on the Internet)? The answer may well be that different customers want different things and that channel organization must accommodate their wishes if it is to maximize sales.

- **Location 'plus':** This separate heading makes the point that there is more to a location than just where it is. A high street may seem convenient, but how close is parking? Travel to the place of purchase is a major factor in customer choice, and customers are especially fickle about such things. In the nearest small town to my home, even the far end of the high street is regarded as less accessible than its core area (and parking is pretty good).

- **Service:** This is important too. Some products need a degree of advice and information to assist their purchase, and all benefit from service in the pure 'good-customer-contact' sense. If your product needs advice before people will choose and buy, then the chosen channel must be able to provide it. Again, there are compromises to be made here – but customers will decide what's best for them, and may well have the option of deciding that your product is just too complicated to buy and moving on to a competitor that has things better organized.

Although it is right to use the word *compromise* in so complex a setting, customer satisfaction is clearly vital and everything else must take account of that.

### 6.6.4 Managing the channels

This heading separates the need to manage the channel mix in an organizational sense from the separate need to influence people along the routes (see the next point). Managing the channels in this sense means:

- incorporating specific **channel-focused activity** into the marketing plan;
- setting **clear priorities** for making each channel work;
- **allocating people** appropriately (for example, who will be accountable for the customers in different channels and how these customers will be serviced – remember the differences inherent in the car-tyre example earlier);
- creating and maintaining appropriate **service backup** for each channel (and for each kind of customer, including the ultimate customer).

It also includes matching a whole range of activities to the nature of particular channels and customers. For example, transport and distribution need to be equipped with trucks whose size relates to the deliveries that need to be made to particular kinds of customer; change this and the fleet may be rendered obsolete.

### 6.6.5 Influencing those involved along the channels

This is certainly one of the most important areas to consider. You have to sell to customers, and create, maintain and develop relationships with them; this links to another topic a little beyond our brief here but worth mentioning, that of account management.

Always, there is a need for communication with distributors to:

- **inform** them about the product, your plans, how things are going, opportunities to come – and more; an ongoing dialogue is necessary here;

- **motivate** them, especially if contact is infrequent or difficult, as with overseas distributors;
- **support** them, with, for example, joint promotions, merchandising or advice, training – anything they expect or find useful.

This ongoing communication may use every available method, from simple email to video conferencing, and must be planned and sustained (and therefore budgeted for in both time and money). It is all too easy to think that, just because your product sells well, a contract exists and your advertising is good, intermediary people will complete the role you hope for with regard to selling on the product unbidden, as it were.

Success in managing channels effectively is down in large measure to this sort of communications programme; it should have clear objectives, be systematically undertaken and be creatively executed.

### 6.6.6  Treat channels like markets

In all respects, this is an important one. Channels are dynamic and can be fast-changing. The people in different markets see things differently and expect their viewpoints to be understood and respected by their suppliers.

All activity needs to be tailored to individual channels. This is easier when channels are very different (as in the earlier car-tyre example), but needs more conscious thought when differences are less obvious.

### 6.6.7  Tailor products and service to channels

A well-known brand of chocolate may be an international brand, and it may well appear essentially the same in Bangkok, Boston and Brussels. In all likelihood it is not. To take just one factor as an example, the product must have a higher melting point when sold in hot countries. Now it might be made possible to sell the standard version in a hot part of the world (it could be transported in refrigerated containers). But it must also survive local distribution, and it might well be that no one will handle it unless it does so – after all, it is the local distributors who will get the complaints.

This is an obvious example perhaps, but it does show how everything must relate to the market and thus to the channel. Instructions are necessary if furniture is to be sold through some channels in flat-pack form. Packaging (or boxes) might contain more information for technical products sold through channels whose outlets can or do provide less advice and information than others, as might instructions (and both will need translating for overseas markets).

Thus an important part of the decision making and management relating to channels is about fitting products to channels, rather than simply fitting channels to (existing) products. More complication, yes, but it helps get a fit that works effectively.

### 6.6.8  Match marketing activity to individual channels

This needs comment in terms of marketing at various levels. Let us take advertising as an example. It may need to:

- incorporate channel information to make it clear how something can be purchased;
- be directed not only at the ultimate consumer but also at intermediaries along the distributive chain.

And it always should be appropriate and well directed to its target audience. Different strategies are involved here. Some international brands use the same advertising worldwide (e.g. Coca-Cola); others will focus on different markets in different ways.

This kind of link has to be borne in mind whatever the product. Once a strategy has been settled on, its implications must be worked out. Look, for example, at the difference in advertising in the computer area between Dell, who sells direct and whose ads target potential purchasers, corporate or not, in a very similar way, and other companies, who need a response that takes people into a dealership that may well sell various competing brands. The latter want people to walk in already feeling strongly that their brand is right, and perhaps with a preconditioned accept-no-substitutes frame of mind.

The range of circumstances here is immense, but the principle is clear. Everything along one chain may differ from everything else along another and all marketing activity must be designed to work in the knowledge that that is so and of what those differences are.

### 6.6.9 Monitor channel performance

A channel is not for life. The dynamic situations inherent here have been mentioned more than once. Given that what is being implemented is a mix, it is a prime part of channel management to monitor performance. To be more specific – to monitor comparative performance of all channels in order to:

- improve the way an individual channel performs;
- better balance the mix, putting more emphasis on one channel and reducing it on another, perhaps;
- ultimately, perhaps, to change radically the way distribution is handled.

The quick and easy way to review this is with the market-map concept mentioned earlier.

For whatever reasons, the overall aspects of channel selection and management seem to remain unreviewed longer than many other aspects of marketing. This is both likely to leave you open to danger and to risk missing opportunities. Review should be a prescribed and structured process.

### 6.6.10 Innovate and seek new solutions

Finally, as part of the ongoing regular review mentioned above, new developments should be observed and new options sought (after all, *someone* has to be first with some things – why not your organization?).

Let us be clear what we mean here by 'new'. It would be entering new territory to organize party-plan selling for a product never sold that way before, but it would not be *new* – the technique has been tried and tested elsewhere. But it could well create a new mix of channels used and that, in turn, might produce increased sales.

Sometimes there is significant innovation that affects many organizations. This may be something like a move to out-of-town shopping centres, or something technologically led such as the whole area of IT and e-commerce. In practice many innovations are seemingly minor changes – evolution is more common than revolution – but that does not mean that they are not worthwhile. They may well be, and a series of them may be still more useful.

## 6.7   An international dimension

If business is international, the principles outlined here are all equally applicable. Additionally it will be important to:

• fit methodology to markets; this may mean a variety of ways across the different countries or territories involved, all of which suit the local culture and ways;
• not allow distance and difficulties such as language to affect adversely the need to support and motivate distributors (whatever form they may take);
• set things up in a way that does not run foul of the local legal system; this may affect the actual contract made with an independent distributor, for instance;
• keep the whole thing manageable; the wider the network and the more organizations and people it involves, the more difficult it is to maintain suitable communications.

Overall however, despite any difficulties involved, expanding the places where a product is sold is a classic option in building the business and this can be done only when based on a clear strategy for distribution.

## 6.8   Summary

What makes channel management successful is much the same as what makes any aspect of marketing successful. Like marketing itself, it needs to be:

• customer-focused, because ultimately, however innovative, however clever something is, the final arbiter is in the marketplace;
• continuous in its implementation and in its review, because ongoing refinement and improvement should be inherent in a dynamic marketplace;
• coordinated, because a complex mix is involved and one aspect of it may well affect all the others;
• creative, because what matters is what works, and preferably what works better than whatever it is competitors are doing, not what the rule book says.

If the principles of good channel management are well applied, it has the potential to contribute positively to the marketing process and, in so doing, to help provide a positive edge in the market.

*Part III*

# Strategy and marketing planning

While the commonsense nature of much of marketing should now be apparent from what has been covered to date, the broad scope and complexity of the process should be, too. Because of this, and because of the fact that marketing is central to the whole process of making an organization successful and profitable, an ad hoc approach is unlikely to maximize effectiveness.

Marketing must be clear about what it is trying to do. It must have a clear, logical and creative strategy that is effectively the route to achieving its objectives. This must be formalized in a plan, not least to specify what will be done (and not done), how, by whom and when. As the marketing activity predicated in the plan proceeds, and as time goes by, progress must be charted to ensure the right results are achieved progressively and that the organization both learns from what is happening and can act to fine-tune activity and performance.

The goal of the next part of the book is to illustrate the options presented by how marketing strategy can be approached and the principles upon which selecting strategies is carried out. Similarly, at the end of this part, readers should understand the essential principles of marketing planning, and also appreciate how the activity such a plan triggers is monitored to assess progress towards objectives and provide a basis for fine-tuning activity if matters do not proceed exactly to plan (which, is incidentally, the norm).

*Chapter 7*

# Marketing strategy

Before marketing activity can be organized and implemented, it must be clear both *what* it is aimed to achieve and *how* it will achieve its aims. Strategy sets out the course of action; it is essentially the way in which action will proceed to achieve results. Strategy can be decided upon only in the light of information – first, in terms of taking practical account of how marketing works overall (set out in Chapters 1 and 2); second, in terms of information about the organization, its product and its markets, which may well come in part from research – and certainly from analysis.

This chapter looks at what is necessary ahead of compiling a marketing plan.

## 7.1  In a position to succeed

The plan, and the marketing planning process, dealt with in detail in the next chapter, pose a number of options; indeed, given free range, the variety of options open to the marketer, at a number of levels, may be considerable. Yet only one course can be pursued at a time, at least in a particular marketing situation, and those in charge have to make sometimes hard decisions.

We will look at this by reviewing in turn three levels of decision making:

- the definition of **the market** (the group of customers/potential customers with whom an organization hopes, or rather intends, to do business; as usual this may be complicated by any international dimension);
- the setting of clear **objectives** (desired results in the chosen marketplace);
- the selection of **strategic direction** (the course of action that is intended to achieve that result).

The above three levels may seem basic and obvious, but it should be noted that they are also fundamentals of success. In any organization, especially one without dramatic product advantages, the clarity and focus that are given to operations can be a major element in any success enjoyed. Thus, companies who expand fast and successfully – Starbucks is a good current example – almost always have a clarity of purpose that gives strength to everything they do both in marketing and those elements that support it (for example, staff motivation in the case of Starbucks).

Consider the three in turn:


## 7.2    The market

The plan must recognize that a definition of market – indeed, the market that a company believes it serves – is often difficult. *Too narrow a definition and a company's alternative strategies can be limited, perhaps disastrously. Too wide a definition and a company's strategies and resources can be diluted across too many competing opportunities.* Neither bodes well for success. The problem is that the concept of 'market' has many interpretations. A market may be, for example:

- a group of people having a common interest (car drivers; computer buffs; vicars; training managers);
- a particular part of the world (Milton Keynes; Granada TV area; the Middle East; the southern sales region; the EU; South East Asia);
- a broad business sector (information technology);
- a narrow business sector (dictionaries, within the total book market);
- a particular area of need that cuts across many market segments (computers).

Deciding which market a company is in (or should be in) is crucial to the objectivity – the focusing – of the corporate plan, and, importantly, of the company's marketing, product management and pricing policies.

Having identified the market segments in which it is competing, a company has to assess how its products, their presentation and their packaging match the needs profiles of those segments chosen. In doing so, the company has to recognize it has the choice of aiming at four broad market types.

1. **Undifferentiated markets:** As the name suggests, this approach provides a blanket, take-it-or-leave-it marketing package that covers just about everyone likely to buy. This strategy (which ignores the fact that consumers make choices and enjoy doing so) can be dangerous, because it leaves a firm open to attack by competitors attempting to seize sub-segments by appealing more closely to very specific segment needs.
2. **Concentrated markets:** This attempts to match closely services or products with the needs of a narrow market segment. Being a big fish in a small pool has many advantages, especially for smaller firms lacking the resources to compete in much broader markets or multimarkets. This specialist kind of approach has its disadvantages, too. If there is a business downturn in one narrow segment, the company may lack a sufficiently broad base of alternative buyers to whom resources can be quickly redirected to create new sales.
3. **Differentiated markets:** Broadly, this represents a compromise between the above two extremes. This marketing approach looks at the entire range of segments within a particular market and seeks to satisfy those of sufficient size and

potential reward with precisely targeted but very similar products. (For example, Volvo Penta manufactures diesel engines for pleasure craft. Within that closely defined market it has a model for every significant segment from tenders to 50-metre luxury cruisers.)

4. **Served markets:** This is a term that implies little purposeful planning, because *served markets* refers simply to the markets that a company presently serves, whether or not there is advantage in so doing. In other words, market opportunities have led the company into sectors perhaps in an uncoordinated manner, rather than the company devising a coordinated strategy to take it into the most advantageous sectors. Such companies are said to be opportunity-led rather than strategy-driven.

With a clear view of the market it is aiming at, the company can turn to setting objectives.

## 7.3    Marketing objectives and strategies

Clear objectives exist to focus and to place any tactical activities in order of priority. *Clear* means quantified wherever possible (rather than all-embracing statements: 'We will aim to make as much money as possible'). A much-quoted mnemonic illustrates whether this has been achieved: objectives should be SMART, that is:

**S:** specific
**M:** measurable
**A:** achievable
**R:** realistic
**T:** timed

More about objectives is explored in the next chapter.

Beyond that, the overall objectives available to succeed and grow in the market must be considered in light of the strategies that will be deployed. There are, in fact, surprisingly few overall objectives – perhaps six main ones, shown here, followed by some possible strategies for achieving them.

1. *To increase the share of the existing market* (necessarily winning business from competition) – through: concentration on selected segments; developing product applications; using different brand names.
2. *To expand existing markets* – through: increasing frequency of customer purchase; increasing usage; opening new branches.
3. *To develop new markets for existing products/services* – through: approaching new market segments; export marketing.
4. *To develop new products/services for existing markets* – through: revision of old products; introduction of radically new things.
5. *To develop new products/services in new markets* – through: diversification; takeover; technological extension – through: improving value offered to customers; marketing audit and productivity improvement and such tactics as reduction of range.

6.  *To increase profitability of existing business* – through: offering improving value to customers; marketing audit and productivity improvement; reduction of range and more.

Such options are not, of course, mutually exclusive. Often, a combination can have an even stronger effect on marketing plans. However, the greatest danger for an organization, at the point of selecting appropriate strategies, is that it may be tempted to adopt too many courses of action. Such a mistake can easily spread management too thinly and prevent commitment to putting maximum effort behind the prime and most important courses of action.

Marketing planning, then, must begin with a thorough and creative attempt to choose the most appropriate strategy for the entire organization's marketing effort. Whichever route is chosen, it is best reviewed against a checklist to see that the strategy:

- satisfies the needs of the various precise target groups at which it is aimed – consumers, wholesalers, customers, etc.;
- achieves the corporate marketing, financial and growth objectives;
- gives direction to various elements of marketing activity – products, prices, distribution, and promotion;
- blends well with any other strategies, i.e. does not hinder their achievement;
- capitalizes on the corporate strengths and minimizes the effect of any weaknesses;
- creates a competitive advantage that is difficult to match or surpass;
- is within the competence and resources of the company.

At the same time, decisions must be made about positioning – the place selected within the range of options represented by the market. For example, Ford and Porsche both make cars, but, although Ford makes some higher-performance cars, it would be seen as somewhat downmarket from Porsche. The two are positioned differently from each other in the market – something, incidentally, that is separate from their size, profitability or other measures that might be made of them.

The nature of a product is usually considered to be the most important element of the total marketing mix. Based on the marketing objectives and strategies of a company, management must make a variety of decisions on product mix, product lines, brands and services. These decisions are critical to the continuing prosperity of the company and it is therefore important to assess fully, relative to competition, how well products perform in terms of the needs they aim to satisfy. In evaluating product performance it must be recognized that products are bought for what they do rather than what they are; that the benefits conferred by a product are created not only from its actual physical features, but also from a whole constellation of other objective and subjective characteristics such as availability, reputation, after-sales service, fit with lifestyle and more.

In many markets, especially but certainly not exclusively the consumer ones, the brand, that is the product name, is crucial in terms of what is called *brand image*. This is the product and all that goes with it. So, the brand image of, say, an airline includes everything, from the service, the fares, the livery, check-in arrangements and behind-the-scenes elements such as maintenance and safety procedures, to the name and the way it is all presented in promotion and advertising. The kind of brand image that will be created needs deciding on. Sometimes this is wide, designed in effect to be all things to all people; sometimes it is very narrow – what is called *niche marketing* – and directed at perhaps smaller and focused segments.

### 7.3.1 Link to product life cycle

Product life cycle was touched on earlier; in selecting strategies the facts associated with this must be borne in mind. The cost of launching a new product is considerable (and the failure rate is high, too, certainly in FSMG markets, with only one in ten new products surviving any length of time – a further demonstration of the fickle nature of markets).

This means that often the first launch uses a *test market*. Instead of launching nationally, it is tried out in just a smaller region, a county or independent TV region (if TV advertising is to be used). This may reduce the risk, but even then there are key considerations to be taken into account.

- **When to launch the product:** If the product is replacing an existing one, should the stocks of the old product be run down? If demand is seasonal and the season is well advanced, should the launch be postponed until next season?
- **Where to launch the product:** It is important to choose an area where rapid acceptance and payback can be achieved rather than launching it in the main stronghold of a competitor.
- **To whom to launch the product:** Often, certain sections of the population are more open to new ideas than others and so are willing to try it. These might often be the first segment targeted, especially with the likes of IT products.
- **How to launch your product:** A clear promotion strategy needs to be planned to get public relations, advertising and selling to reinforce each other and produce optimum results.

What has been said here so far suggests clearly both the need for a clear path of action and the complexities involved in creating a clear strategy. There are various approaches that can assist the process of analysis and decision.

## 7.4 Strategy: planning and analysis

A variety of approaches exist that act to assist in focusing the mind on the way through the complexity. Their role is to organize and arrange matters so that good decisions can be made and the best possible results made more likely. It should be borne in mind that there is no one right plan or strategy. All that can be done is to ensure the decision is made on a well-considered basis and that as many pertinent facts as possible are in mind as this is done. The best-known and perhaps most useful formal approach here is the ubiquitous SWOTs analysis.

## 7.4.1    The SWOTs concept

This acronym stands for *strengths* and *weaknesses*, *opportunities* and *threats*. Essentially it only formalizes a commonsense view of what needs to be investigated early in the planning process before strategies are set. It is a classic means to an end. SWOTs analysis prompts the questions that should sensibly be asked at this stage. These may need investigation before they can be satisfactorily answered, and the answers may need some analysis before it is clear what they are actually saying.

Diligently addressed, the process ensures that planning can proceed on the basis of a sound knowledge of the situation *within* the organization (SW) and *externally* (OT). Some answers will be obvious and well known. Others will not, and investigation may produce surprises. Even one key area of sound information being flagged may affect action materially.

*Note:* If such an analysis has never been done in the past it will take some time. But on an annual basis (planning usually being an annual cycle), it is not a daunting part of the total process to update this picture; and certainly it is a useful one.

That said, here SWOTs is perhaps best explained further through an example of the sorts of question that are used during the analysis (which can, of course, be applied to products or services in any sort of organization).

## 7.4.2    An organization's strengths and weaknesses

### 7.4.2.1    A: Customer base

A.1    What is our current customer base, by size, by location, by category?

A.2    How does our disposition of customers (customer mix) compare with the market mix?

A.3    Are our customers in growth sectors of the market?

A.4    How (as a specific measurement) dependent are we on our largest customers?

### 7.4.2.2    B: Range of services

B.1    How closely does our product range reflect the market's needs?

B.2    How does our range compare with competitors'?

B.3    Are the majority of our areas of business in growth or decline?

B.4    Is the span of our product range too narrow to satisfy our markets?

B.5    Or is our product range too broad to allow satisfactory management of performance across the range?

### 7.4.2.3    C: Price structure

C.1    What is the basis of our pricing policy?

C.2    Do our direct and indirect competitors structure in the same way?

C.3    Are our prices competitive?

C.4    Do our customers perceive our prices as offering value for money?

### 7.4.2.4    D: Promotional and selling activities

D.1    With which customer groups are we communicating?

D.2 What do they know and feel about us?

D.3 Are we communicating with enough of the 'right' people (both groups and individuals)?

D.4 What means of communication are we using?

D.5 What attitudes exist internally that influence approaches to promotion and selling?

D.6 Is each person in contact with customers capable of selling our full range, and doing so equally well for every element of it?

D.7 Do they possess the necessary knowledge and skill for selling?

### 7.4.2.5   E: Planning marketing activity

E.1 Do we have agreed plans for marketing and selling?

E.2 Do the plans state specific activities as well as objectives and budgets? And are they measurable and able to be monitored?

E.3 Do we have individual/departmental as well as corporate plans?

### 7.4.2.6   F: Organizing for marketing

F.1 How is the firm's marketing activity organized and coordinated?

F.2 Are authority and responsibility for each person/activity clearly defined?

F.3 Are all our people committed to contributing to a marketing culture that will assist the achievement of commercial success?

### 7.4.2.7   G: Control and measurement of marketing

G.1 Have we defined 'success' for our staff and ourselves?

G.2 Have we established all the necessary key result areas to measure that success?

G.3 Do these standards examine marketing as well as publishing goals and standards?

G.4 Do we measure performance against desired standards and take appropriate (and prompt) corrective action?

## 7.4.3   *Market opportunities and threats*

### 7.4.3.1   H: How is the market structured quantitatively?

H.1 How many people/organizations of what type are there in our market with a need for our kind of product?

H.2 What are their current buying practices?

H.3 How much do they spend on such items?

H.4 How often do they buy (e.g. annually/monthly)?

H.5 Whom do they buy from currently?

H.6 What do they not buy?

H.7 How do existing and potential buyers access our market and our kinds of product?

### 7.4.3.2    I: How is the market structured qualitatively?

I.1    Why do existing and potential customers buy/not buy?
I.2    What do they think of what they buy (e.g. good value/overpriced)?
I.3    What do they think of those who supply their current needs (e.g. too big/too small/helpful/unhelpful)?

### 7.4.3.3    J: How is the market served competitively?

J.1    Who are our direct competitors (i.e. other similar companies)?
J.2    Who are our indirect competitors (i.e. 'overlapping' companies, some of which it may be easy to overlook as uncompetitive)?
J.3    What are their strengths and weaknesses (list their size, staff, image, pricing, marketing skills, geographic coverage, etc.)?

### 7.4.3.4    What are the quantitative and qualitative trends?

- market/segment size
- market/segment requirements
- market/segment structure
- market/segment location
- competition

Such an analysis is an invaluable tool in charting a course into the future. Which areas are most important will vary depending on the size and type of organization concerned. It may seem to indicate a daunting task, and indeed it can be a significant one, but once it has been addressed it needs updating and carrying forward rather than necessitating starting again with a blank sheet of paper. This is part of the concept of what is called a *rolling plan*, one that builds on the past as it moves on to the next year and beyond – five, ten or more years (the Japanese have companies with one-hundred-year plans).

### 7.4.4    *Practical use of SWOTs*

The questions in the example above form a starting point. Certainly they are indicative of the sort of thinking that must be done, and of the sort of information that needs to be clear to allow effective operation to follow. Every company needs to tailor its approach – and its specific questions – to this kind of thinking, in terms of the nature of both the company and the markets in which it operates. For example, customers can be defined very differently. If an organization sells direct or only to one industry, then the power of the retail chains may not concern you. But all the specifics of a particular mode of operation must be accommodated.

There are, of course, further planning aids. By way of example, the Product Portfolio Analysis of the Boston Consulting Group is described in Panel 7.1. Beyond that, others you might see used or want to investigate include the Ansoff matrix and the GE matrix.

## Panel 7.1 Boston grid

The Product Portfolio Analysis developed by the Boston Consulting Group is a much-copied strategic tool that reviews a product's market growth rate and its market share, presenting a graphic picture that can assist strategic planning. The following shows the general concept. The categories of product used are defined broadly as follows.

**Stars:** As the name suggests, these are products that are doing well. They have a strong market share and good prospects to retain or grow it.

**Cash cows:** These are currently good, they have a substantial market share now, but their prospects for growth are low; meanwhile, they can produce significant profit.

**Dogs:** These are bad in both respects: they have low market shares and are rated low too in terms of growth prospects; frequently the situation in mature markets and candidates to be phased out.

**Problem children (or question marks):** These are products with a low current market share, with prospects but needing a substantial investment of time and money to gain real growth; might become either stars or dogs.

Such a technique is, in part, just to focus, though, alternatively, it can be used with some precision and the graphic picture may have the members of a product range positioned exactly within their appropriate square.

### 7.4.5 Other approaches

It should be acknowledged that there are many prescribed approaches (or permutations of them) that, in one way or another, help with the definition of strategy. They are not mutually exclusive, have different focuses and all assist the process of seeing through the complexity in a way that makes sense of things and allows practical, action-oriented decisions to be made. In recent years many of these are tied to particular management gurus and writers. A good example is Porter's five-forces model. This takes the view that it is competitive pressures that primarily shape an organization's strategy; he commends viewing external matters under five headings:

- threat of new entrants
- bargaining power of buyers
- threat of substitute products
- bargaining power of suppliers
- industry (current) competitors

Again this clearly concentrates thought, though no such simple model encompasses everything; others suggest that Porter ignores human-resources aspects. In reality, any organization must select and use a manageable suite of techniques to ensure that they do not disappear into endless navel contemplation; and SWOTs may be all that is needed.

Incidentally, much of the best writing and comment on the core element of marketing planning has affected best practice significantly. Without a doubt the main and best-known guru here – and one who has a great many sensible things to say – is Philip Kotler. His seminal work is *Marketing Management* (Prentice Hall) and there is much in this voluminous textbook about planning and more about strategy. He has a list of books to his credit, some with a specific industry focus. A recent book, briefer than some, which gives a quick overview of his thinking, is *Kotler on Marketing*.

## 7.5    The scope of planning

The description 'business plan' has an air of substance about it. What do you plan for? You plan for the business, of course, and marketing plans and strategies are a major part of that. But it is also important to consider the level down to which planning is taken, and the most important aspect of this is planning at the level of individual customer. Customers are subject to the 80/20 rule (Pareto's Law, which, in this context, says that 80 per cent of your revenue and profit is likely to come from around 20 per cent of your customer list). This means that large customers are different from smaller ones and that, while small customers are important, major customers warrant some significant individual attention. If planning and strategy seem unnecessary at this level, consider the effect of losing even one key customer. Size here has no specific measure. It is what is important in a particular case and thus, in a small business, may be of a size that a large company would rate as small.

To illustrate what is worthwhile here, consider two approaches. The first is designed to flag up opportunities in major customers and help plan sales strategies to exploit them. The second addresses the topic of customer profitability.

### 7.5.1    Strategic planning for an individual customer

Again, a matrix approach is useful here: one that plots revenue from sales of different products alongside 'buying points' (these might be different locations, people or departments across a large organization). In my training business, for example, different departments might buy different courses: sales by sales, business writing by a central HR department and so on. If the significant turnover from a big customer is split across the matrix, gaps may show up. There may be good reason for these, or they may represent an opportunity and need a change of sales tactics that can be linked into, and specified in, the plan. Figure 7.1 shows how this works.

### 7.5.2    Planning to protect and increase customer profitability

This has come to be more important as customers have polarized and the big have got bigger. As customer demands or services offered (or both) extend, more and more margin is vulnerable to being eaten up in just getting the business.

Any company that analyses the costs of obtaining business may be shocked at just how many things seemingly conspire to reduce margin. These include all the costs of

Buying points

*Figure 7.1    Sales maximization matrix*

the sales force (from recruitment to commission); discounts (and there may be many different bases for them, e.g. quantity bought or when purchase is made); special packing; delivery (maybe to multiple locations); labelling; credit terms; advertising and promotional support; merchandising assistance; training of customers' staff ... The list goes on. Unless this is addressed carefully profitability may well be in danger. A plan for large customers may be useful, but the situation should be measured quantitatively with individual customers of appropriate size, to make a link to an action plan for anywhere the figures demand action is taken.

*Action:* management must plan by addressing the list of possible 'profit diluters', the policy involved and sales staff's attitudes and competence. For example, it may be that policy is at fault and profit being lost because published terms need attention, or that policy is right, but that salespeople are losing out due to poor negotiating skills.

Overall, while there are a variety of techniques to consider, what you do must be a practical compromise that allows planning to take place and reflects the needs of *your* business. The example that follows (in Panel 7.2) shows how radically planning (and the thinking it entails) can change an organization's nature.

## Panel 7.2    How planning can create positive change

To illustrate that what matters most is the thinking process that is inherent in all this, consider how something as seemingly basic as asking, 'What business are we in?' can affect the direction taken. The view that the question is answered by may change progressively but is useful; it not only needs keeping up to date but also, in the longer term, as it is thought through, can affect the whole direction of the business. An example is perhaps the best way to make this clear.

Company X were a scaffolding firm, a traditional business; asked what business they were in, they would originally have replied – seeing it no doubt as self-evident – 'Scaffolding for the building industry'. True enough, though by its nature a current or historic description and a somewhat limiting one.

As they looked ahead, and searched for markets beyond building, it was discovered that other areas of construction (as varied as chemical plants and North Sea oil rigs) needed scaffolding. The description shifted to 'scaffolding for the construction industry', and the organization changed to take advantage of these additional opportunities.

So far, so good. But what was it they really sold? Discussion defined this as being 'temporary access and support', now a little way from the original description of the business, and this led to further searches for market opportunities and inroads being made with the leisure industry where large amounts of scaffolding are now used in contexts such as temporary spectator stands at everything from pop concerts to sporting events and parades.

Seeking to take this thinking still further, and finding that technically their systems (joints, fixing, etc.) were little different from those of competitors, they concluded that they did have an edge in *the skill with which they were able to erect and dismantle their scaffolding, in terms of speed and safety*. While the company had never seen export markets as a possibility (steel tube is prohibitively expensive to transport overseas), expertise can be exported. This, after further work (it is not being suggested that these kinds of shift are achieved without time, effort and investment), led to good business being developed in training others (for instance, in the Middle East) to erect and dismantle scaffolding to the same high standards.

The change wrought by this kind of thinking is clear; and all that is needed is to make time for a little constructive thought. Setting strategies and planning are thus a long way from academic (in the negative sense of that word).

## 7.6  Summary

A huge range of approaches and techniques can be deployed here. The important things are that:

- there should be clear objectives, since without them there is no basis for deciding any strategy;
- the chosen strategy should be well considered and based on as many objective facts about the business as possible;
- analysis should take place as a core part of the thinking (whichever and however many techniques may be used to focus it);
- final decisions about strategy should be based first and foremost on a market view (and certainly not just on what is convenient to the organization);
- all this thinking should dovetail neatly into a formal and systematic planning process (the subject of the next chapter);
- the essentially creative basis of marketing should not be forgotten; there is no way of arriving definitively at some one 'right' answer in terms of the way forward, and sometimes success is as much with those whose planning and setting of strategies unashamedly include some true inspiration.

*Chapter 8*

# Marketing planning

An old maxim in business says that, if you do not know where you are going, any road will do. Certainly, given its complexities, marketing cannot be left to chance. Here we investigate the process involved in drawing up a marketing plan. Realistically, the scale of things varies here a great deal. A small business might put its plan on one sheet of paper, or not have one (though the latter is not recommended); a large business may need plans that set out matters in relationship to a range of different products and their intended performance in many different, perhaps global, markets. What is set out here goes down the middle, as it were, in terms of scale, but it is intended to touch on the key issues.

## 8.1 Why plan?

### 8.1.1 The core reasons

It has been said that planning is only anticipating the inevitable and then taking the credit for it. In reality, it is much more than that, of course, and the (good) marketing plan is a key practical element in creating business success. Marketing planning and the plan it creates are a creative process, based on information and analysis that prescribe the plan of action that is designed to execute marketing strategy, and produce the business results targeted.

Given the dynamic nature and complexities of business, sound planning is essential and nowhere is this truer than when applied to marketing. Marketing has so many options for action; not least the marketing plan selects those actions that will be taken (bearing in mind that, as resources are always limited, no organization can do everything).

In this chapter we review the art of what is a key management task: actually creating a plan, doing it well and making sure it is of practical use to the business.

The effective business plan must recognize and balance:

- the needs of the company, the needs of its staff, and of other groups such as shareholders;
- the demands of the external environment and the market (customers);
- the activities of competition;
- the resources and capabilities of the organization.

The plan for the year – the annual plan – may well be an integral part of longer-term planning with the operational plan for the immediate future linking to outline plans, and lines of thought, for the longer term. The marketing plan, a core element of the business plan, has very specific purposes, spelt out in the list below.

To be of practical benefit (and justify the time taken in preparing it), a good marketing plan will:

- identify opportunities for future profit improvement;
- have the ability to anticipate dynamic external changes;
- provide better protection for the future of the business;
- prompt the collection of relevant data;
- allocate the company's resources towards specific ends;
- underpin the process of control;
- assist with clear communications around the company;
- focus individual efforts and assist personal motivation;
- provide a proper commercial reference for all activities;
- justify development (and development funds).

Further, if its preparation is to be a practical process, it will help if:

- the approach is an integration of 'bottom up/top down' (i.e. it involves people throughout the company);
- the system and purpose are clear to all;
- standard (tailored) planning formats are used;
- a planning cycle, specifying all timings, is agreed (and starts well in advance);
- the planning includes a facility to 'fine-tune' (particularly to take advantage of opportunities);
- an eye is kept on the external reactions to everything done to facilitate 'fine-tuning'.

Someone must take responsibility for the planning process (and give some time to it), while others (including you?) must agree to be involved as necessary. Some discipline may be required here, since other pressures too easily intrude. In the context of this work, while all aspects of this planning are important, the promotional aspects – the things that will bring in the business – are key. Marketing, in the guise of the marketing director or manager, must be responsible; if no such specialist is available (in a small company) the burden falls on the general manager or managing director.

The starting point should be having a clear overall view of the business.

## *8.1.2   The specifics of the marketing plan*

Examining the specific content of the marketing plan also helps illustrate why it is important and what it can do. Any revenue can come *only* from *outside* an organization, so how markets are addressed is paramount. The plan overview must specify the core elements it should contain and the key issues it should address. As an example, the marketing plan should usually include:

- a statement of basic assumptions regarding likely future developments (for example, in short/long-term economic, technological and social changes);
- a review of the past sales (revenue and profit) in as much detail of individual product, market, geography and even major customers as seems useful;
- a statement of the external part of SWOTs – the opportunities and threats;
- an analysis of the organization's strengths and weaknesses (the internal side of SWOTs), in terms of such factors as facilities, human resources and skills, finances and customer franchise – also parallel competitor information is useful here, so far as it can be ascertained;
- a statement of long-term objectives and the strategies for achieving them;
- a detailed statement of the objectives and strategies for the year ahead – taking this to whatever level is appropriate: i.e. individual product, market and elements of marketing activity;
- a look ahead at how the plan for the next year will need to pick up from that for this (and for years beyond that, depending on the organization's scale of operation);
- a statement of priorities showing what is key to the plan and how the organization will capitalize on its opportunities, identify and correct any weaknesses and so on.

The plan needs to take the form of a specific *action* plan, scheduling what will be done and by whom, and the full timing implications of implementing the plan. This needs to be so for all the component parts of marketing utilized (that is, public relations, advertising and promotion, sales, etc., and the detail of each). It is here above all that activity needs to link tightly to the budgets and financial statements. A plan without an action focus is never so likely to succeed.

With this much in mind we turn towards the process that assembles it, and what makes a good starting point.

## 8.2   Mission statements

A mission statement is a now ubiquitous piece of jargon, and a prerequisite early part of marketing planning. It describes what is simply a succinct but all-embracing statement about a business and its role, purpose and goals.

It is not so much *having* a mission statement that is important, though it does have considerable merit in communication terms, acting to enthuse employees, customers, shareholders if appropriate, with the feeling that the company knows what it is doing. Rather it is *being able to construct one* that is vital. For, without thinking through much about the organization, writing such a description may be impossible, and, if you are not clear in overall terms about what you are trying to achieve, how can you ever devise a detailed plan? Mission statements are not panaceas, but they *do act* as useful means to an end; they are influenced by and influence the marketing side of the business. Preparing one is a logical first step in the overall planning process. They can be devalued by being turned into a public-relations statement; this can result in something nice-sounding but less useful for planning purposes. Specifically mission statements should:

- define the kind of business the organization is in;
- identify dimensions of business that current plans exclude;
- focus on customers, specific customer categories and customer benefits;
- link to benefits to stakeholders (e.g. shareholders, owners and, not least, employees);
- say something about the organization's culture and values (these may be an important part of the profile of the firm, as with Bodyshop's environmental attitudes).

One important point here: a mission statement does not reflect a solely *internal* view: it must describe the organization primarily in terms of its *outside* involvements with markets and customers.

## 8.3   The core planning principles

Essentially, marketing planning stems from five key questions. These may seem deceptively simple, but between them they define the process, illustrate its logic and dictate what must be done to accomplish it. They are as follows.

**1. Where are we now?**
This needs addressing through research and analysis – and is sometimes called *'situation analysis'*.

**2. Where are we likely to be (at a particular future time)?**
This is the situation addressed by forecasting, whether this constitutes an intelligent guesstimate, or sophisticated statistical techniques such as regression analysis.

**3. Where do we want to be?**
This is simply the objectives: what you want to achieve and by when.

**4. How will we get there?**
This relates specifically to the objectives you set – and specifies the strategies you will use to get there; it is the action plan.

**5. How will we know whether we are on track and – ultimately – when we have got there?**
This reflects the needs for management control and control systems. It also reflects real life: no plan can be a wholly accurate description of what will happen. It is designed to be as accurate as possible and, if you do achieve it then that is good. But a

plan is essentially more like a route map than a straitjacket. It must not restrict you to one course of action, but allow – and help – you to take action to change and fine-tune things as a period goes by. You may plan your ideal route on that route map, but it will also help you find alternatives if things do not go exactly to plan (if your chosen route has roadworks, for instance).

Now, we address the actual planning process. It needs a clear, well-defined and systematic approach. The next series of headings are therefore subsections of the overall approach, which encapsulate those elements currently seen as the main issues. This is intended to be a good, practical guide – and, not least, to demonstrate a manageable approach. In real life, though core aspects are common, the detail of what any individual needs to do must be tailored to the situation of their own organization.

## 8.4   Compiling the marketing plan

### 8.4.1   First steps and analysis

Naturally, writing a plan presumes that those doing the planning know the markets in which they operate, have clear marketing objectives and have the power to authorize or recommend action to agreed cost levels. Each aspect commented on here will correspond to a section of the written plan, the extent of each varying with the needs of the company.

All companies should have financial goals expressed in budgets of revenue and expenditure (in the overall business plan) and the first task is to note them.

However, since sales and profits can be made only from customers in the markets we select, the next task must be to translate the financial objectives into market objectives. These must answer the question, 'What results must be achieved in the marketplace to produce the financial objective required?'.

Only by considering two interrelated analyses can meaningful answers be produced.

- What opportunities and threats will be present?
- What are the strengths and weaknesses of the company?

These together form the SWOTs analysis (strengths and weaknesses, opportunities and threats), which was dealt with in Chapter 7. This only means taking a hard, objective look inside the company and out, before setting down more of the plan; it may be worth rereading the description of this, which sets out the kinds of question that need to be asked, in the last chapter alongside the current commentary.

For the planner, there is more here than can reliably be kept in mind and thinking it through and making some notes that may influence action are valuable.

### 8.4.2   Objectives and strategies

The next step is to translate financial goals into marketing objectives that bear in mind the SWOTs analysis, then, with the objectives in mind, to identify strategies that could be pursued. It is worth repeating, from the last chapter, the precise difference between *objectives* and *strategies* and the precise definition involved here is crucial.

- **Objectives** are the desired result sought from the marketplace.
- **Strategy** is a course of action designed and intended to achieve planned results.

Let us take these in turn. Without objectives, it is impossible to focus and to place any tactical activities in order of priority. Yet, the main options available to us in marketing objectives are limited, primarily the six listed in the previous chapter.

From the analysis of market opportunities and threats and the internal assessment of strengths and weaknesses, we can select the marketing objective(s) that will best achieve the financial goals for the planning period.

Next, objectives must be linked to strategies, the purpose of the strategy being to focus effort, coordinate action and exploit the identified strengths of the organization. By corollary, the purpose is also to avoid wasting resources on peripheral and nonproductive activities.

Clearly, different objectives will require very different strategies, and the end result must be a set of well-matched objectives and strategies; again, see the last chapter.

Remember, the selection of strategies need not be mutually exclusive. Often a combination that combines several different strategies can provide an even stronger effect in marketing plans. However, the greatest danger at the point of selecting appropriate strategies is that we may be tempted to adopt too many courses of action. Such a mistake spreads management, and other resources, too thinly and prevents commitment of maximum effort to the prime and most important courses of action.

Marketing planning must continue with a thorough and creative attempt to choose the most appropriate focus of the entire organization's marketing effort. The determination to concentrate simplifies the tactical marketing plans, which must then follow for the range of products/services to be offered, the prices to be charged and the promotional and selling actions to communicate with the chosen markets.

### 8.4.3   Products and price

There are factors to be considered here such as product life cycle and positioning, which have been dealt with elsewhere. However, we do need something that will prompt ongoing thinking about both. Products need to be changed or updated and must keep up with the market; services, too. And price must be watched constantly for profitability and competitiveness; such decisions must not become merely formulae.

With products/services, consider particularly that:

- long-established, mature products/services may remain the same in principle as they were years ago, but they will have to change in detail to meet changing customer needs;
- new products/services may have to be developed, both to meet new customer needs and to keep out competitors who see them as a means of ultimately acquiring more business;

- there may well be opportunities to offer a range of differently priced variants for different customers and different situations;
- as any harsh economic climate forces corporate customers to seek productivity improvements in all functions of the business, so they become more demanding, but the same pressures may also produce new opportunities.

And with prices ask yourself these questions.

- How aware are customers of price – the actual levels or such measures as hourly rates (e.g. car service)?
- How do customers perceive price? Does, for example, a higher price imply in their minds higher quality or do they view price as a commodity factor with no differentiation between firms?
- Are there 'price barriers' in a customer's mind, which we must avoid in any quotation for business (e.g. £100 or £10)? How far can we price differentially because of the perceived and accepted reputation we have?

Price is a critical area of the marketing mix and one that tends to receive too little analytical attention. Far too often, the decision is simply to keep in line with the competition or to work essentially on a cost-plus basis. In fact, price should reflect the overall policy at the strategic level and show creative flexibility at the tactical level, up or down, depending on the threat or opportunity.

## 8.5 The promotional plan

This is given its own section because a number of different elements are involved. Successful promotional activity needs to be based on a continual process of review and action, and preparing and implementing a comprehensive promotional strategy demand time, skill and a systematic approach. This is an inherently important part of the plan and also makes a good example of the detail into which different sections of the plan must go.

### 8.5.1 A systematic approach to the promotional area

The checklist below makes clear 12 key planning tasks, which will then be examined in more detail.

1. Analyse the market and clearly identify the exact need.
2. Ensure the need is real and not imaginary and that support is necessary.
3. Establish that the tactics you intend adopting are likely to be the most cost-effective.
4. Define clear and precise objectives.
5. Analyse the tactics available, taking into consideration the key factors regarding:
   - the market
   - the target audience
   - the products/services offered
   - the firm's organization/resources

6.   Select the mix of tactics to use.
7.   Check your budget to ensure funds are available.
8.   Prepare a written operation plan.
9.   Discuss and agree the operation plan with all concerned and obtain management decision to proceed.
10.  Communicate the details of the campaign to those involved in implementing it and ensure that they fully understand what they must do and when.
11.  Implement campaign, ensuring continuous feedback of necessary information for monitoring performance.
12.  Analyse the results, showing exactly what has happened, what factors affected the results (if any) and how much activities cost.

To simplify somewhat, these can be considered under five main sequential stage headings.

### 8.5.1.1   Analyse the company's needs

The prime difficulty in the analytical stage is not so much the identification of the need, but ensuring that the need is real and not *imaginary*.

Identification of a need can come from:

- formal, and informal, research
- internal company investigation
- staff
- specific market demands
- personal observations

Such analysis is part of the total marketing review (and SWOTs). Promotionally, we are primarily concerned to show clearly the interrelationship of customer categories (i.e. the kind of firm/organization/individual they are), products/services and business (i.e. new business – a new customer, extension – an existing customer buying more etc.). We may well have to plan different strategies to impact specific areas, for example to produce sales to a specific kind of business with whom we have not had prior contact.

Once a need has been clearly identified, it must be established that the kind of support you intend using is likely to be the most cost-effective method of fulfilling that need. Then the planning stage can commence.

### 8.5.1.2   Preparing the operation plan

The first stage of any plan must be the quantification of the objectives. A clear statement is needed of exactly what you want to achieve, stated as specifically as possible. Listing an objective stating, 'To improve business' is just not precise enough. Whereas an objective that states, 'To improve the number of new customers buying Product A by 50 per cent this year' makes it clear to everyone exactly what needs to be done and, above all, how success will be measured.

Once the objective is finalized the selection of tactics can take place. This will depend on a number of factors, including the following.

**The market available:**

- What is its nature?
- Is it buoyant or is it in a low period?
- Is it price-conscious? If so, how?
- What is the competition doing?
- What is the customer profile?

**The target audience:**

- Types of people/organization?
- What are their buying habits?
- What motivates buyers?
- What are their current attitudes to promotion?

**The products/services offered:**

- What is our current performance?
- What are the strengths and weaknesses?
- What promotional support has it received in the past?
- Capacity available?
- Market profile/image?
- Position in life cycle (i.e. is it seen as new and interesting or old and dull)?

**Organization of the firm:**

- What are our current sales and promotional methods?
- Would some tactics cause internal difficulties, e.g. in terms of administration or resources?
- Is the company involved in any other activity that might affect what we want to do or detract from it?

With these questions answered, there may still be a number of alternative tactics, all of which could be suitable for achieving the objective. Which tactic to select and use may then depend on which is the most cost-effective.

Once the decision on tactics has been made, the details should be formalized into a *written* operation plan. It is always worth writing this down, even in a small firm. This should not be a one-off exercise, but will eventually provide a reference, which can be updated regularly so that it always sets out the plan for the next period. Planning of this nature is a rolling process. It should include:

- background information as to why the promotional support is necessary;
- the objectives;
- profile of the target audience(s);

- reference to product/service details;
- details of additional support other than that which you are actually planning, perhaps that being done by associated offices, or various staff;
- budget details – how much the action is estimated to cost;
- details showing exactly how the plan will be implemented;
- controls, standards and methods of obtaining results;
- an action plan, or timetable, showing what actions are required, when they should be carried out and by whom.

There are a variety of ways of making the decision on the budget more logical, e.g. using comparisons with competitors, standard percentages of revenue and so on.

### 8.5.1.3   Preparing for implementation

As long as the operation plan has been correctly prepared, the pre-implementation preparation should be a formality. This can be achieved only if the operation plan has been discussed and agreed with everyone concerned with the support activity, well before any action is required. This can ensure you pick up ideas (or identify snags) from everyone in the firm, some of whom may surprise you with their constructive comments.

Do not forget, either, that if everybody feels involved they will more readily commit themselves to the next stage.

### 8.5.1.4   Implementation

The success or failure of any promotional activity, provided it has been thoroughly planned, then rests on how well it is implemented. The effectiveness of the implementation will depend on how well the details are communicated around the company and then controlled. Therefore, the details of what is to be done must be communicated in such a way that they are clearly understood by everyone.

Effective methods of controlling the implementation must be set up to obtain maximum feedback while promotional activity is running. This will permit any necessary changes to be made at the earliest opportunity.

### 8.5.1.5   The analysis of results

Any promotional campaign can involve a great deal of personnel time and is often expensive in terms of opportunity cost. This is true regardless of what is spent on the other aspects. You therefore want to know how money is being spent and what achievements are obtained from that expenditure. Examining the detailed results of every form of promotional activity will show clearly:

- what the situation was prior to the activity;
- what we aimed to achieve (the objective);
- what the situation is after the promotional activity has ended (and what we have achieved);

- whether there are any factors outside our control that might have influenced the result, what they were (e.g. competitive activity, legislation changes) and their effect;
- what has happened to the rest of the market or at least our near competitors;
- what the effect might have been had we not carried out the promotion;
- what the budget was and how it was spent.

Careful analysis of what has been achieved is important, not least as part of the planning and consideration of what to do next, which should be occurring in a continuing cycle.

## 8.5.2   Beyond promotion

No promotional activity plan can be carried out in isolation, particularly without linked sales follow-up and service along the way. This must be planned, too, so, as an addendum to the promotion plan, there must be a focus on sales activity. Here we should think about and list who will do what.

- How much prospecting will be done, when, how, by whom?
- Who will follow up leads, to what timescale?
- What sales targets are necessary?
- What record will be kept?

Planning and implementing a soundly based systematic promotional plan are not easy, nor is ensuring that all the backup resources, people, skills and systems are geared to converting the initial enthusiasm created in potential customers into actual business. But it is certainly necessary and, done successfully, it provides a sound basis for securing and, more importantly, enlarging, the business.

Most people plan and work with the promotional plan best if it is, in part, in the form of a calendar; indeed, in a small business this might constitute most of the plan! This creates an ideal rolling plan (a perpetual year planner will give you a 12-month view at all times) and the detail will fill out as time goes by. Thus, in January, 99 per cent of the activity will be listed for, say, January to March, whereas only half might be able to be listed at that point for the next part of the year. Using a wall chart, with plenty of space, tailoring the headings to your business will provide a practical planning and implementation tool and also shows the relationship in time terms between different activities at a glance.

Finally, it may be worth having a section to deal with 'other issues'. This flags the implication of the rest of the plan and links to other elements of the business. Here are some examples.

- Does a sales initiative (involving formal presentations, perhaps) necessitate training being done in advance?
- Does recruitment need to build in any new skills required in future?
- Does the organization or structure need changing (e.g. a new job description)?
- Do systems/controls need adjustment?

As has been mentioned any – and every significant – activity area may need its own plan as a subsection of the overall plan. Panel 8.1 lists possible areas where the marketing plan may pose issues of a more general nature that will need attention in any overall business planning or flagging to those running other functions.

---

### Panel 8.1 Marketing/business plan interface

Links from marketing to other functions and areas might include (in no particular order):

- management roles, structure and succession
- legal matters
- research & development and new product launch
- action to alter relationship with competitors
- production, quality control and capital equipment
- personnel and HR activity
- training
- information technology
- design

If these, and more, need additional documentation – and liaison – then so be it.

---

Business planning, along with all its components, is a broad issue. The foregoing represents a minimum approach to the core marketing part of such a plan; it should not put you off doing more (and investigating more), but think carefully before omitting the thought or action implied in any of the areas that are specified. A sound foundation makes everything in this book more likely to hold up.

There can be merit in keeping plans for the existing products and new ventures separate (while, of course, ensuring that they operate in an integrated fashion); if that is done, then new ventures may justify the addition of further dedicated plans.

## 8.6   An element of control

Finally, never forget that one way in which marketing plans assist is with control. Two things are important here.

- **Corrective action:** If actual results diverge from plans, then controls need a *diagnostic* element. It may be necessary to discover why things are going wrong and to fine-tune activity so that the deviation is overcome and activity, albeit revised, still produces what the plan intended
- **Positive feedback:** Just as important is analysis when things are proceeding *better* than planned. Again, questions need asking and if useful lessons are learned there may be positive lessons to build into future plans, or positive action to take fast.

This point is expanded on separately in Chapter 9.

## 8.7 Summary

Details not withstanding, the key issues here can be summarized in just a few words. A plan must be:

- approached systematically;
- based on sound information;
- practical and able to assist positively the decision making and *action* that drive the business.

To ensure this, a practical approach must be taken to the process. Ten points conveniently summarize the management process involved in creating a marketing plan; it must be:

1. given some time;
2. formalized, i.e. linked to agreed formats, timing and other internal systems;
3. based on up-to-date data;
4. subject to some analysis (whatever techniques may be used to assist with this);
5. discussed: getting inputs and ideas from all the appropriate people is part of the process;
6. finalized and agreed;
7. documented: a plan must, by definition, be in writing;
8. communicated: to all those in the organization who can benefit from knowing the plan and who may be involved in any way in its implementation (a cut-down version of the plan may be needed for this);
9. linked to implementation;
10. the basis for a rolling plan, i.e. one plan becomes a major input for the next.

*Chapter 9*

# Coordination and control

## 9.1 Organizing the marketing function

As we have seen, in order for an organization to achieve its objectives, some form of planning system is needed. This, in turn, demands an organization structure both to create marketing plans and to implement them.

For a marketing-oriented organization (that is, focused on consumer needs), the structure should be built from the bottom up by looking first at customers and their needs. Whatever the form of marketing organization adopted, care must be given to its integration with other functions of the company. Then marketing activity can proceed and the next consideration is the monitoring and control of what occurs.

Control in marketing is conceptually similar to other forms of management control. It depends upon the comparison of *actual* (A) performance against preset *standards* (S) and the taking of corrective action based on the resulting variances (V) in accordance with the formula:

$$A - S = +/- V$$

The objectives and plans will form the basis of the standards. For instance, the sales plan will contain sales forecasts, which can be translated into targets for each member of the field sales team.

Because there are so many and disparate influences on market results, many of the promotion areas have been traditionally viewed as largely immeasurable. There are, however, exceptions to this. For example, responses to direct mail promotion can be tracked very accurately, especially if this is the only form of promotion used; so too can hits on a website. Undoubtedly, it is difficult to assess and control activities that, for the most part, depend on human reaction and cannot be easily separated from other influences. The difficulty has been exacerbated, however, by attempts to evaluate each part in terms of the whole. Thus, 'Exactly how much will advertising affect sales?' is usually an unanswerable question, because sales do not depend upon advertising alone.

*Table 9.1   Planning elements*

| Concepts | Illustrations | Contents |
|---|---|---|
| Objectives | The corporate *destination* | What needs to be achieved. This must be expressed by objectives that are:<br>S ... Specific<br>M ... Measurable<br>A ... Achievable<br>R ... Realistic<br>T ... Timed |
| Strategies | The *road* the company will travel to reach its destination | A description of how the company will achieve its objectives:<br>S ... Segments<br>A ... Audience<br>P ... Product/features/benefits |
| Tactics | The *vehicle* used to carry the company along its strategic road | A description of:<br>P ... People responsible<br>A ... Actions to be completed<br>M ... Methods to be employed |

If, however, it is clearly specified what the advertising is supposed to do – perhaps to impress a certain message on a certain number of people of a certain type – then it is feasible to define standards of performance and evaluate achievement. Certainly, control of a different kind can and should be exercised in marketing. All marketing activities are quantifiable in money terms and thus budgetary control can be exercised.

For marketing activities to be effective it is not enough that each department be well run. The whole marketing function needs to be integrated and well coordinated with the rest of the company.

## 9.2   Implementing and controlling the plan

### 9.2.1   The planning elements

A marketing plan, deriving from a SWOTs analysis and the other processes referred to in Chapters 7 and 8, should spell out three key elements of the activity that it plans. These are: objectives, strategies and tactics; these are illustrated in Table 9.1

Successful implementation depends upon the effective management in five key areas:

- **What:** the goals, objectives, aims;
- **Who:** the people responsible/accountable for plan realization;

- **Where:** the specific, identified, quantified marketplaces;
- **When:** the period covered by the plan;
- **Why:** the reasons, desired outcomes.

It will be perhaps a marketing manager's responsibility to ensure that all material marketing resources are in place. These could well include such as:

- product
- packaging
- internal promotion aids (e.g. sales aids)
- external promotion campaigns (e.g. advertising)
- budgetary management systems
- sales data systems
- performance standards

Once the plan is launched, the emphasis of attention changes to control.

### 9.2.2   Exercising control

The plan should contain specific financial, marketing, sales, distribution and promotion objectives. These provide a basis for performance standards in a number of key areas, including product sales (in value terms), product sales (in volume terms) and, where appropriate, in market-share terms. Performance standards can be expressed as follows.

- **Annual targets:** These express performance expectations in sales, profit and market-share terms. Annual targets are known as *absolute targets*. Variances to these targets will identify what has gone right or wrong, but not why.
- **Moving standards:** These express the annual targets in moving divisions of the plan period, i.e. monthly or quarterly actual and with cumulative figures and trends. Again, although moving standards can forecast deviations from plan, they will not identify why performance is greater or less than the required targets.
- **Diagnostic standards:** These can identify what is causing the variations and why, and may indicate an appropriate action.

Table 9.2 amplifies these points.

### 9.2.3   Variance analysis

Variances are calculated by comparing actual results against the preset standards. First, cumulative totals can be used so that individual monthly variations will tend to cancel each other out. Second, moving annual totals (MATs) can be used by taking 12 months' performance up to and including the month in question.

As each most recent month is added and the same month of the previous year is deducted, the trend in moving annual total will indicate present performance compared with the same period in the previous year. This enables comparisons to be made on a single diagram of monthly performance against target, cumulative performance

*Table 9.2    Marketing plan control*

| Type of control | Objective of control | Standards | How to measure performance | What to look for examples |
|---|---|---|---|---|
| 1. Annual product plan | To examine whether plan objectives are being achieved | Sales quotas and market-share financial targets | Comparison of actual results against standards set in each area of performance | Notable shortfall between standard and actual; failure of individual sales territories to achieve sales targets by buyer category |
| 2. Profitability | To examine whether financial objectives are being met | Profitability by product or product group | Comparison of actual results against standards | Major shift in production mix; spending levels above plain levels; declining sales |
| 3. Efficiency/ productivity | To evaluate and improve results of marketing expenditures | Promotional deadlines. Distribution targets. Sales force activities | Comparison of actual results against advertising plan; sales force: who called on/how many/call frequency/ what done in each call | Failure to meet deadlines or set standards in each area of promotion |

against target and, via the moving annual total, the present year versus the previous year.

The benefit of a good control system and feedback mechanism is that they enable management quickly to *identify sales performance variances and the true reasons for them, and to react to changed circumstances*. By monitoring the plan they will be in a position to report monthly (or at whatever frequency is necessary), and answer the following questions, which may be raised by top management.

- Are the plan objectives being met?
- What are the variances between budget and actual?
- What are the causes of these variances?
- What actions are being taken to correct them?
- Is a re-forecast/re-budget necessary?

Marketing does not guarantee success but, carefully applied, it certainly increases the likelihood. In a commercial environment, an organization's success is measured by its profitability. To be clear about just how this is defined, the next section sets out further detail.

## 9.3   The profit mechanism

### 9.3.1   Measuring corporate performance

The best overall measurement of corporate performance is usually taken to be return on capital employed (R/CE). This ratio is calculated by expressing net profit (usually before tax) (R) as a percentage of capital employed (CE). Capital employed is calculated by taking the total assets less the current liabilities, which is of course equivalent to the fixed liabilities, the long-term money. For those with even a smidgen of doubt about their numeracy, key financial terms are defined in Panel 9.1.

---

**Panel 9.1   Ten key financial terms defined**

**Return:** Net profit (usually before tax and sometimes before interest charges where levied by a parent company on a subsidiary).

**Capital employed:** Fixed assets plus working capital.

**Fixed assets:** Money tied up in land, buildings, plant and machinery.

**Current assets:** Money tied up in stock, work in progress, debtors, cash, etc.

**Current liabilities:** Money owed in the short term, e.g. creditors, overdraft, tax, dividends.

**Working capital:** Current assets minus current liabilities.

**Fixed liabilities:** The long-term money in the company, i.e. shareholders' equity, long-term loans, retained profits.

**Margin:** Sales revenue minus cost of goods sold.

**Balance sheet:** The overall statement of a company's position at one moment in time, usually at the end of the financial year, showing the sources of the money in the company (fixed liabilities, current liabilities) and how it is deployed (fixed assets, current assets).

**Profit-and-loss account:** A statement showing the results of a company's trading in the period, indicating revenue, expenditure and thus profit.

This primary ratio of R/CE is a function of two secondary ratios: return (R) on sales (S) multiplied by turnover of the capital employed (CE) and the number of times the capital employed (CE) is utilized in generating sales revenue (S). Thus, the basic profit mechanism of the company can be represented as:

$$\frac{R = R \times S}{CE = S \times CE}$$

This equation, in turn, depends upon other ratios. A change in return on sales must be due to a movement in one or any combination of the following four factors:

- sales volume
- price
- cost
- product mix

For example, a potentially dangerous future position can easily be masked if figures such as sales volume and revenue are viewed in isolation. If volume increases have been achieved by price cutting or revenue expanded by selling more of the low-margin products from a range, then ratio analysis will help identify the concomitant risk to profitability.

A change in capital turnover will be caused by either a rise or fall in the utilization of fixed assets and/or working capital. Furthermore, for example, if the turnover of working capital has altered, it must be as a result of relative change in the relationships of current assets and/or current liabilities to sales. Analysis of capital utilization in this way can help reveal that the apparently successful sales campaign has created such a build-up in debtors that profitability has suffered.

This process of reasoning can develop a ratio hierarchy down to departmental detail, which will analyse total company operations. Figure 9.1 shows a typical ratio chart.

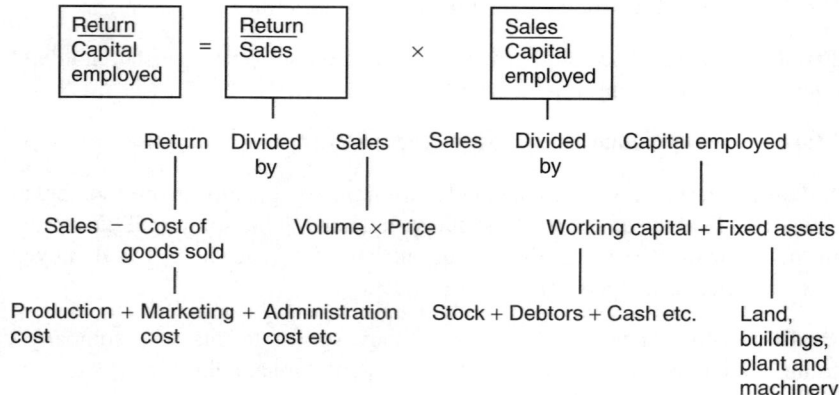

*Figure 9.1   A typical ratio chart*

## 9.3.2 *Identifying the profit input of marketing*

A ratio tree covers total company operations and can be used to identify marketing's contribution to profit making. Most sales and marketing people will recognize their responsibilities in the R/S portion of the equation; this is a natural development from the sales-oriented approach. In this half of the hierarchy, the ratios can be employed very effectively to analyse the reasons for performance change. It is usually illuminating to identify the relative importance of volume, price, cost and mix changes, particularly in those organizations that seem to believe that the solution to every profitability problem is just to increase volume.

Some problems are inherent here. For example, it is surprising how few sales managers clearly recognize their inputs into the S/CE part of the mechanism.

As Figure 9.1 shows, two of the major components of working capital are stocks and debtors. Although a sales manager may not directly control credit granting, debt collection and stock availability, their policies must surely have an impact on them and thence on the profitability of the company. Interestingly, there is a paradox here. If the sales manager wishes to increase sales volume, two of the tactics that they might well use are immediate ex-stock delivery and generous credit terms. While such inducements will almost certainly increase sales and probably the R/S half of the equation, they could have a disastrous effect on the sales/capital-employed area, leading ultimately to lower profitability in terms of the ratio R/CE. All of this reminds me of being told by my sales manager in my first job, 'Remember – it's not a sale until the money is in the bank'. Wise advice.

This quick analysis of the marketing input to the profit mechanism immediately shows up areas of responsibility that need to be defined.

- Who should set stock levels?
- Who should control credit when it is used as a promotional tool?
- Is the product mix sufficiently specified and controlled?

It should be possible in every company to identify which managers are to be responsible for which ratios. That alone is a major step forward in planning and controlling the business, if only because it highlights the interfunctional conflicts that must be resolved. For example, if the production manager is wholly responsible for finished stock levels they will tend to keep them low to save cost, whereas, if the sales manager is given that authority, they may well decide to increase both the range and the depth of the stocks in order to capitalize without delay on every sales opportunity. Usually some compromise must be reached in the light of the impact on the overall profitability.

The control element of marketing is evident in many ways. Consider one example, specific to e-marketing, of which many people have experience. If you buy books (this one, even) from the Amazon website, then every time you log on you can see your buying (and search) record being linked to recommendations. In this way, literally mouse click by mouse click, the customer information recorded is turned into tactical marketing activity to build the customer relationship and promote new sales. The immediacy here is peculiar to e-commerce, but the principle of using past information to prompt action to influence the future is classic.

### 9.3.3    Ongoing systematic review

So finally here we will end on a checklist note, because ultimately what makes marketing succeed is attention to detail, constant review and fine-tuning of action against a background of the factors mentioned above and elsewhere. Systematically reviewing all the key areas of marketing, and the things that affect them, ensures that marketing is not only set up in a way that will make success more likely, but stays organized for maximum effectiveness. The questions that follow, therefore (which touch on some areas still to be covered in detail), are what the marketing should be investigating on an ongoing basis.

Logically we start by looking externally.

#### 9.3.3.1    Environmental factors

**Economic/political:**

- What is affecting our market and customers?
- What is affecting our costs?
- What are the implications for our pricing policy?
- Are changes to corporate policy necessary?

**Environmental** (a more recent and increasingly important area):

- Are there dangers to image?
- Are there opportunities arising from prevailing thinking?

**Social/legislative:**

- Is legislation affecting promotion, price, product or any other area of marketing activity?
- Are changes going to affect our staff or customers?

**Technology:**

- How will the various technological changes inherent in the modern world affect products, customers and how we operate?

The net result of all these questions is to ascertain whether the company's philosophy, culture, planning and organization are appropriate to the situation in which it must work now and is likely to have to work in future.

#### 9.3.3.2    Market factors

- Are customer needs changing?
- Is the market changing in nature?
- Is the customer group aimed at changing in structure or location?
- Is the demographic basis of the market changing?

### 9.3.3.3 Distributive factors

- Are ways of accessing the market changing?
- Are there changes in the area of physical distribution?
- Are the channels and intermediaries through whom business is done changing in nature or requirements?

### 9.3.3.4 Competitors

- Is enough known about direct competitors?
- What about indirect competition?
- Are new competitive pressures on the horizon?

### 9.3.3.5 Marketing activity

This is not, as we have seen, static. It needs constant review and change. So questions must be asked of current practice on a continuous basis in the following areas.

**Customer attitudes:**

- Are potential customers well identified?
- Are their needs understood?
- How well do current offerings satisfy those needs?
- What image do they have of the company?
- What view do they take of competitors?

**Corporate intention:**

- Is the relationship between marketing and required profit clear?
- Is financial performance measured against competitors?
- Is management able to monitor financial performance in a way that monitors progress and enables both corrective action to be taken and opportunities grasped?

**Product/service:**

- Is the status of the product reviewed regularly?
- Is its position in its life cycle understood?
- Is product range satisfactory in extent?
- Is a product-development system in place?

**Price:**

- Is price market related?
- Is pricing policy reviewed regularly?
- Is pricing structure (terms, discounts, etc.) right?

**Distribution:**

- Does current policy give the access required?
- Are outlets managed, dealt with and motivated effectively?

- Is the physical distribution method cost-effective?
- Is distribution regarded as a variable just like other marketing variables?

Next, the various promotional techniques must be constantly monitored to ascertain how they are performing.

**Public relations:**

- Who should be communicated with?
- What do they know of the company already?
- What image should be presented?
- What feedback is regularly obtained?

**Advertising:**

- Who should be communicated with?
- Are messages appropriate and persuasive?
- Is what is done objective-oriented rather than fashion?

**Sales promotion:**

- Is it clear at which points of the marketing process sales promotion is appropriate?
- Is what is done organized towards specific goals?
- Is promotion organized as an integral part of the mix?

**Selling:**

- Is the detail of the sales job well defined?
- Are all methods of personal contact being used cost-effectively (e.g. teleselling)?
- Are the salespeople well trained and well managed?
- Is the sales team appropriately deployed?

**The mix:**

- Are all aspects of the promotional and sales mix well integrated?
- Are sufficient time and effort being put into planning and organizing all of this compared with implementing activity?

### 9.3.3.6   Corporate matters

Everything in marketing works better if it takes place against a background in which the company overall is clear in its intentions.

- Is it clearly defined what business it is in?
- Are profit goals spelled out in terms that relate to the various levels and divisions of the organization?

- Is there a plan that sets out objectives and specifies the implications in terms of specific, timed action plans for all those involved?

In addition, there are three further areas of review that are corporate in scope and need clarity and regular review.

**Marketing planning:**

- Is there a plan?
- Is there a satisfactory system for producing and updating the plan?
- Are the right people involved in the process?
- Does any necessary documentation assist rather than restrict the thinking involved?
- Is the whole process constructive and does it prompt creativity?
- Is the plan communicated? Indeed, does it stimulate constructive communication around the organization?
- Is the plan regularly evaluated against subsequent events?
- Does it link to individual action and assessment?
- Does it work as a true operating document (rather than gathering dust on the shelf)?

**Marketing organization:**

- Is everyone involved in marketing organized so as to maximize effectiveness?
- Does the organizational structure ensure the action that is required happens promptly and constructively?
- Are individual jobs well defined and responsibilities understood?
- Is everything in marketing based on sound experience and training?
- Are the supporting systems arranged to help the activity or does administration apply an unintentional brake on what is being done?
- Is information necessary for marketing available, up to date and appropriately and promptly circulated?
- Are communications around the firm as they should be, both within any marketing department and more widely around the company so that *everyone* involved in whatever way understands how they assist the marketing activity and results?
- Is motivation an asset to marketing initiative (or a drain)?

**Marketing control:**

- Is what constitutes success well defined?
- Are measurement devices in place?
- Are all cause–effect factors understood and monitored?
- Are management information and management information systems sufficiently market-based?
- Is there a focus on key results areas?

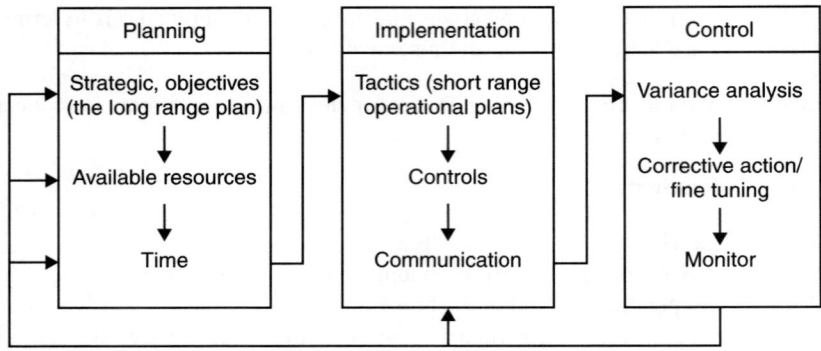

*Figure 9.2    Implementing and controlling the marketing plan*

## 9.4   Summary

Remember the four key questions posed earlier in discussing planning:

- Where are we now?
- Where do we want to go?
- How will we get there?
- How will we know when we get there?

Simply put, control answers the final question, and not only answers it but allows opportunity for fine-tuning of performance. It provides a hand on the tiller to adjust course when things are not going exactly to plan and, just as important, the possibility to take advantage of changes that present, sometimes unexpected, opportunities. Thus, with this in mind, we see how important coordination is to the marketing process. The progression from planning to implementation to control in a continuing cycle (illustrated in Figure 9.2) is as important overall as any of the individual elements; together they make the process work. With all of this, good factual information is the basis of the best decisions, and any source of good information is to be welcomed.

Now we can move on and look at communication out from an organization to its market and customers.

*Part IV*

# Marketing communications

The third P of the marketing mix is presentation (or promotion). After all that has been reviewed to date, there is a prime ongoing need to tell people – actual and potential customers, and others – about the product and the organization that offers it. 'Telling' is a somewhat inadequate description: actually a variety of communications and techniques are involved, each directed in different ways at different target audiences with a variety, and scale, of objectives in mind. While many different types and styles of communication are involved here, the overall intention is that these should be orchestrated into a total communications strategy to execute the most effective – and cost-effective – promotion.

The goal here is, first, to comment on the basis of all promotional activity, which is the way people buy and how they make buying decisions. Promotional and sales communication needs to reflect how this happens. Beyond that readers will see what the various techniques within the promotional mix are, how they each fulfil their separate individual roles and how they work together to create a total promotional strategy.

In addition, these techniques – advertising, sales promotion, public relations and more – are linked to the role of personal selling, the customer-service aspects of business and the building of long-term relationships and customer loyalty. Something of the blend of creative and practical principles that can be utilized to make this area of marketing effective is also reviewed, and so is the ever-present need to differentiate from competition that makes finding new approaches a constant consideration of the way marketing works.

*Chapter 10*

# How people buy

Marketing is, however unfortunately, associated with inappropriate pressure – indeed, even unsavoury or actually criminal tactics – to make people buy. While some people trade on the fact that 'there is one born every minute', this is a small, exceptional area and not what mainstream marketing is about. Not least because most organizations want to do regular business with their customers, they recognize that sales made and found inappropriate by customers tend not to repeat. To build business with customers – satisfied customers who may well return for more – demands respect for the customer. More than that is involved: the whole mechanism of promotional and sales communication must relate to and use the psychology of how customers assess what is offered to them and make a decision to buy or not, and to buy from one supplier rather than another. Let us consider the implications of this.

## 10.1   Types of buyer

It must be recognized that people may be, at any time, at different stages in the buying process, and marketing communication has the challenge of communicating effectively with each. The pyramidal diagram (Figure 10.1) illustrates the different types of buyer. As can be seen the marketer commences with a mass audience, then progresses upwards towards a decreasing – but all-important – number of customers. Many factors will influence whether a suspect converts to a customer and remains a loyal customer. These issues must be considered in relation to the factors that influence buyer decision making illustrated later in this chapter.

Consider each of the categories in Figure 10.1 in turn.

### 10.1.1   Suspects

A suspect is a person, or group of people, within the total population that the marketer has reason to believe could be interested in the product or service on offer. They have

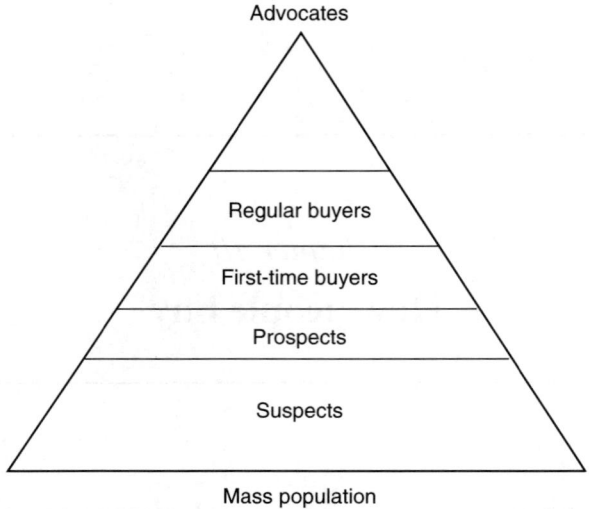

*Figure 10.1    Types of buyer*

not previously enquired or bought the product/service. For the marketer they are an unidentified audience. One way a marketer could target this group is to place an advertisement in magazines of which they are known to be readers.

### 10.1.2    Prospects – identifying target audiences

A prospect is an individual who has not yet bought the marketer's product or service, but has indicated an 'interest'. For example, having seen an advertisement, an individual might respond by requesting further information. The prospect may be gathering information from several sources to help them in their personal decision-making process. They may be currently in a position to buy or may be considering a purchase sometime in the future. Either way, how the marketing team then reacts to carry this interest forward may make the difference between a purchase and no purchase.

### 10.1.3    First-time buyer/customer – converting enquiries to sales

Here the customer has made their first purchase. This may be their only purchase of a particular product or service, or they may become a loyal long-term customer. Progress may depend upon how the organization develops the relationship with the customer; and that, in turn, may be dependent upon a mixture of the customer's experience of the product and service, and whether the organization takes a short-term or long-term view.

### 10.1.4   Repeat buyer/customer – winning repeat orders

Here the customer has made more than one purchase, and might show every indication of perhaps becoming a long-term buyer. The relationship between the customer and the organization has the potential to develop – to be made to develop. Several factors will influence the customer as to whether they make repeat purchases in the future. These will include the following.

- Some products may actually be one-off purchases, e.g. as with the purchase of a house. In some cases another house is not purchased, and the existing one may be handed down through several generations of the family.
- There may be a long time span between purchases, as with, for example, the buying of a new car. An individual may be loyal to a particular brand; however, they may not choose to trade in the old model for a new one for a period of several (perhaps many) years.
- The relationship between the customer and the organization. If a positive relationship is developing then the customer may, depending upon their own personal circumstances, increase their purchasing. For example, visiting a particular restaurant on an increasingly frequent basis because the food (product) and staff attitude (service) are to their liking.
- Some products create the opportunity for constant repeat purchases. However, other factors impact here. These will include the quality of the product/delivery of the service, pricing and packaging.

### 10.1.5   Advocates – helping to build brand loyalty over a longer time period

These are customers who are loyal to any product or service and purchase on a regular basis (bearing in mind the points raised above in relation to frequency of purchase). Additionally, they tend to recommend strongly the product/service to their friends and colleagues. Examples could include a restaurant, an airline, oil and ball bearings.

## 10.2   The buying decision process

Customers engage in a process of decision making that prompts the decision they make. Here the buyer undergoes a series of steps in order to formulate their final decision. These steps can be illustrated as follows.

### 10.2.1   The recognition of an unsatisfied need/want/desire

The customer recognizes a need and subsequently takes action. For example, a person may develop a headache while out shopping and seek to remedy the discomfort as soon as possible by purchasing a painkiller.

This leads to …

## 10.2.2   *The level of involvement*

The customer determines how much time and effort they will invest in satisfying their need. If we remain with our example, the customer will probably want to find a quick remedy for two possible reasons: to reduce the pain and discomfort effectively and be able to continue shopping, either out of need or desire. Therefore, they will limit the level of the time and effort spent assessing the situation and making a decision (in the jargon, the level of what is called 'involvement').

This leads to …

## 10.2.3   *Identification of the alternatives*

The customer will consider the various alternatives available to them. In our example, they will probably identify that there are several types of analgesic available – aspirin, paracetamol, ibuprofen and more. These are available either through well-known branded names or the own-label brands of a chain such as Boots.

This in turn leads to …

## 10.2.4   *Evaluation of the alternatives*

The customer will analyse the potential advantages and disadvantages of the various alternatives identified. In our example, the customer may consider the strengths available, whether they are allergic to any of the ingredients, any possible side effects, whether they are in tablet or capsule form, the number of tablets/capsules contained in a pack and the price. This may take only a moment, though for other, more complex, purchases it may take considerable time.

This then leads to …

## 10.2.5   *Purchasing decision*

The customer decides whether or not to make a purchase. In our example, the customer may decide to buy own-branded paracetamol capsules because of 'the strength, being easy to swallow and the overall value for money'.

And this leads to …

## 10.2.6   *Post-purchase behaviour*

The customer seeks reassurance that they have made a correct decision. In our example, the quick relief from the pain and discomfort of the headache would provide such reassurance (and affect future purchase decisions).

In reality, consumers must also take a practical view of the possibilities. There are all sorts of things people may feel they want, or even need, but about which the feeling cannot prompt an instant purchase. Most people can elect to buy a headache pill when necessary; few who fancy a Porsche can indulge their fantasy; and, of course, many things need a considered decision taking into account, not least, other priorities. Thus, a decision to buy a new suit or go out for a meal must be balanced against other, perhaps more necessary, purchases. Given the purchasing power that exists, certainly in the developed world, as much marketing is directed at less necessary goods as it is at more essential products.

Another example: we are all familiar with the sorts of goods that are displayed alongside the cash points of a supermarket – confectionery items for one. These are sited there, in part, specifically because the decision time needed before a customer makes up their mind is limited. They sell only from that position because the store recognizes facts about how buying decisions are made.

Consumers will use a set of criteria in order to compare and contrast the different alternatives. The criteria may be determined or influenced by various factors. In Table 10.1 are listed some of those factors, related to the example of the shopper with the headache.

Taking books as an example again, consider the cover of a popular novel. This links tightly with the principles set out in Table 10.1. Past experience is evoked by comparisons with other titles and reminders of the author's past work. The term *bestseller* hints at large numbers of satisfied customers, and references to advertising

*Table 10.1   Possible buying criteria*

| Criterion | Influencing factors |
| --- | --- |
| Past experience of using a particular brand or brands | The consumer may have used particular brands in the past and believe that one was more effective than others. (In reality they may have been composed of exactly the same ingredients, but the person perceives that one brand was more effective than another.) |
| Past experience of using a particular product | In our example the shopper had a choice between aspirin, paracetamol and ibuprofen. They may have discovered that, for example, aspirin irritates their stomach so they try to avoid it. |
| The experience of friends and family | In a previous discussion a family member or friend may have recommended a particular brand of analgesic for its effective action. Therefore, this information may influence the evaluation. Equally, the advice may be not to buy a particular brand because it had little or no effect. With this information the consumer may immediately discount this branded item without any further evaluation. |
| The influence of advertising/marketing | The consumer may have seen a television advertisement for a new brand of analgesic. Therefore, they will seek out this brand at the pharmacy, read the packaging and evaluate it alongside the other brands. |
| Specialist advice | The consumer may seek some form of guidance within the store. In our example they may ask the pharmacist for their advice, even though the final decision rests with them. Equally, the person may be taking other medicines where their doctor has advised that taking aspirin as well may induce mild, but unpleasant, side effects. Therefore they will seek to avoid such additional discomfort. |

or programme tie-ins – 'as seen on TV' – aim to add credibility. Even specialist advice may be there in the form of a review from an authoritative source (the IET 'stamp' on this volume is a case in point). How is something as straightforward as a book marketed in a way that gains advantage from the way people buy?

The whole look is important; so too are other factors. For example, a book appears as a hardback (and at what customers might see as a high price), later as a paperback aiming at a larger market. In addition it may sell through specialist outlets such as book clubs, where either a special price or the need to maintain commitments to the club may prompt purchases. Even the time of year a book is published may be used – a book suitable for present giving being published ahead of Christmas, or a 'holiday read' title coming out in paperback early in the summer.

## 10.3    Purchasing decision criteria

Once evaluation of the alternatives is complete the consumer has to make the decision as to whether to purchase or not. A decision not to purchase may be only a temporary one until other alternatives become available.

In many situations, once the decision to purchase is taken, the time span between the decision and acquisition is often relatively small. In our example of the shopper with the headache, they hand over the packet and the money to the salesperson at the checkout counter – a simple and fast transaction.

However, in other situations the decision to purchase will lead to a series of additional, interrelated, purchase-related decisions. For instance, consider the example of a couple purchasing something more expensive and complicated (the two go together in extending the decision-making process), such as a new washing machine.

The couple may be loyal to a particular electrical retailer, and have thus decided to place their purchase with it. On the other hand, they may have no loyalty to any one store. There is also the factor of experience within particular purchasing environments. For example, the couple in our example might prefer to go to their local high street or shopping mall electrical retailers rather than a department store.

In a retail context, someone's decision to buy at a particular store may be based on a variety of influences; see Table 10.2 for an example of the kinds of factor involved.

Overall there are a variety of influences on purchasing behaviour; some are shown in Table 10.3. There is certainly enough here to suggest that complex behaviour is involved in this area, indeed it is a subject of ongoing research and a host of theories and principles underlie it.

Books also make a good example of hype – word of mouth combined with good marketing that can take a product to great success (witness *The Da Vinci Code*). Overall, the range of influences is broad and disparate. Think about the influences of, for example:

- *distress purchases:* if the car has a puncture and must have a new tyre, then such is usually bought promptly whatever else was originally budgeted for that time;
- *health concerns:* changed risks prompt changed purchase patterns (special socks bought by airline passengers after the deep-vein thrombosis scare);

*Table 10.2     Examples of purchasing-decision criteria*

| Criterion | Description |
| --- | --- |
| The level of helpfulness of the sales staff | In the pre-decision phase, if the staff of a particular store show that they are knowledgeable, have spent time with the couple answering their questions, have been courteous and nothing has been too much trouble, then the couple may be influenced to purchase from that particular store. This is the role of people within the marketing mix (see Chapter 1). |
| The layout of the store | The display and range of products on show may be influencing factors. |
| Discounts/sales offers | If there is strong local competition, some stores may seek to discount the product on a regular basis, or have frequent special offers (including such things as free delivery or installation). |
| Additional charges | Some retailers may charge for a local delivery and fitting, while others do not. This could be a deciding factor; more so if the couple are price-sensitive. |
| Methods of payment | In addition to cheques and credit cards, the store may offer an interest-free monthly direct-debit payment system. This would allow the couple to spread their payments, say over a six- or twelve-month period, at no extra cost. Again, this may be beneficial to the couple if they are particularly price-sensitive. |
| After-sales service | In addition to the standard guarantee/warranty, the retailer may offer a special after-sales service, including a telephone helpline/support service, and the ability to extend the guarantee/warranty beyond the statutory period. |

- *fashion:* how much is bought in the name of fashion? – the much-used phrase 'fashion victim' suggests the power of this factor.

You may well be able to extend this list to reflect your organization or area of the engineering world. For all the theory and the complex mix of influences here, let us consider two principles that link all this to what is done in persuasive communication.

## 10.4    Influencing people towards purchase

Given that ongoing communication is necessary to prompt sales, how do we move the customer, who may start by being totally unaware of our product, to changing their attitude along the way, to being a regular user? There are many techniques we can use, each one bringing us closer to the customer and moving them towards the final goal of their becoming regular users. We can use public relations, advertising, direct mail, sales promotion and many other techniques to achieve this (more of which in the next chapter).

*Table 10.3    Categories of buyer behaviour*

| | Category | Description |
|---|---|---|
| 1. | People | We are influenced in many ways by a variety of different people. They may be our friends, our family, our heroes (for example, football players or pop stars), our teachers, coworkers and more. |
| 2. | Culture | Cultures vary throughout the world. To a greater or lesser degree they influence our lives. The cultural perspective may stem from a belief in one's own country or through religious teachings, as well as simpler factors linked to the kind of circles we mix in and the habits and beliefs of that group. |
| 3. | Lifestyle | Lifestyles can be both real and aspirational. We may be contented with our lifestyle and purchase products and services to fit it. This may include fashion items from clothes to furniture. Equally, we can be aspirational, slowly developing our lifestyle to meet our aspiration of how we really want to live. This can include buying into new product areas as the process progresses. |
| 4. | Economics/ financial | Clearly, our economic and financial circumstances can dictate our buying behaviour. Someone who is unemployed and living on state benefits will be extremely price-sensitive and must focus on purchasing basic products. On the other hand, a highly successful stockbroker in London or New York is likely to be much less price-sensitive, and able to purchase a wider range of products or services. These purchases may include luxury items, a sports car, yacht or who knows what? |
| 5. | Media | The media can have tremendous influence on what we buy. This can range from reviews of films, plays and books (which may be either positive or negative) through to healthy-eating campaigns. The influence of the media can be both a negative and a positive influence: negative publicity can literally kill a product; positive media coverage is very potent and, coupled with other influences, can lead to sales escalation. |
| 6. | Necessities | This covers the basic items that we need to continue living, such as food, heat and shelter. |
| 7. | General/misc. | These cover the influencers that fall outside the above categories. For example, the influences of the government and the weather. |

## 10.4.1   Objectives on the way to purchase

There may be various objectives set for this persuasion, for example getting potential customers to:

- try a product;
- buy more or buy more frequently;

- extend the use of the product thus buying more (as with breakfast cereal as a supper snack);
- instil trust in the company through one brand so that others in the range are also bought; this is important where one brand name – say Mars – is used not just for one product, Mars bars, but also for a range (including Mars ice cream and others).

In publishing, one example is that promotion will be put behind an author as an action that can result in sales of all their titles. Many more sub-objectives might be listed here, all slightly different.

### 10.4.2   Customers: building awareness

The sequence of the customer moving from being totally unaware of your company or product to being a regular user is worth looking at in some detail; it provides a simple working framework against which to consider promotional activity. Each step represents a change in attitude by the customer or prospect. The steps are to move from:

1. **Unawareness to awareness:** This is the stage where the prospective customer moves from no knowledge of a product to knowing about it, or at least of its existence. The prospective customer's attitude is receptive but passive and their major need is information. Promotion is targeted at:
   - introducing a concept;
   - telling the prospect that something more specific exists;
   - creating an automatic association between the needs of the prospect and the product.
2. **Awareness to interest:** This is a move from a passive stage to an active stage of attention. The prospect will have their interest aroused by the product's newness, its appearance or the concept or what is said about it. Their response can be active or passive. Promotional objectives are to:
   - gain their attention through the message;
   - create interest in the product;
   - provide a succinct summary of all the relative information in the prospect's mind.
   (At this stage all the aspects of the promotion mix begin to get the prospect to say, 'I must check this product out'.)
3. **Interest to evaluation:** The buyer will first consider the effect of the product on their personal motivation, which will include image, lifestyle, circumstances, needs and more. They will then go through a process of reasoning, through analysing the arguments and looking for personal advantages. Depending on their needs and price, they may look for more information or other validation of their initial impression. The prospect is now giving out buying signals so an attempt can be made to:
   - create a situation that encourages the prospect to start this phase of reasoning;

- discover and focus on the prospect's relevant wants and needs;
- segment and target buyers according to their requirements.

4. **Evaluation to trial:** This is a key movement from a mental state of evaluation to the positive action of trial. The prospect/buyer's basic requirement is to try the product and evaluate the findings. Promotional objectives are:
   - clearly identify the usage opportunities;
   - suggest timescale for trial.

   (In other words, the intention is to encourage the first purchase and get the buyer to say, 'This does look good and does all I require, I'll buy it'.)

5. **Trial to usage:** The buyer will take this step if the trial has been successful; sometimes Steps 4 and 5 come together. The objectives of promotion are:
   - provide reminders of key elements, such as brand, image or technical advantages;
   - emphasize the success and satisfaction;
   - remind the buyer of other usage opportunities and provide supporting proof via third-party references.

6. **Usage to repeat usage:** This is the final objective of the promotion and the buyer may well be also going through the above stages with other suppliers. When a buyer moves from occasional usage to constant usage, they will have moved into a state of mind of identifying themselves with the product, and selection of the product will be automatic until something happens to upset that balance. The objectives are now simpler, though not necessarily easier to achieve, and are to:
   - maintain the climate that has led to satisfaction;
   - keep up the image;
   - maintain contact and confirm the key qualities of the product, advise of other products as they come along.

The above sequence dissects the process in detail. A buyer having perhaps changed from a product they have used for many years may well feel they are now duty-bound to try several other products, not just the one, now their buying pattern has been upset. So marketers must be on their guard. The customer will probably not be aware of the sequence they have just gone through, and may make very rapid progress from Stage 1 to Stage 6, but the sequence is real, and is what promotion must influence, if necessary doing so one step at a time.

The promotional mix of all the elements that have influenced the customer along the way can be many and varied, from something small such as a leaflet through the door to a major advertising poster campaign on hoardings. With this in mind, marketing can utilize and deploy the separate and individual techniques, making them all work to achieve exactly what is wanted. It is important to bear in mind, when implementing these techniques, that the effect of each is different, and that it is difficult to separate each individual impact. A person's image of an organization is the total cumulative effect of everything they see and hear about it.

That said, consider examples of advertising that you are aware of, and ask yourself which bit of this chain of changing feelings it is directed at. Sometimes this is easily spotted – for example, most advertising for a major brand such as Coca-Cola is directed at past customers to keep them buying and away from competition, rather than explaining the product from scratch to potential new users.

## 10.5 Protecting people's buying

While, for the most part, people try to make sensible buying decisions and understand sometimes when they do not, expecting less, for instance, of some impulse buys, they do not want to be seriously disappointed. In part, confidence and certainty come from the knowledge people have of suppliers – one reason branding works is that buyers regard it as some sort of guarantee. But this feeling can be rendered useless by the unscrupulous.

It is possible for people to make bad buying decisions because they are misled or, worse, actually conned. Given the volume of goods and services sold the incidence of such is minor; just enough to give programmes such as BBC's *Watchdog* the raw material they need. In fact, additional confidence in buying and control on rogue traders comes as a result of considerable legislation designed to protect consumers.

Most suppliers probably want to act in a reasonable way, but the law is there to guide any who are tempted to be less than honest and helps them by describing an environment in which there is control and people can assume this as part of their decision-making consideration.

### 10.5.1 The letter of the law

This is not the place for a legal treatise, but the following highlight main aspects of legislation that affects this area at least in the UK (though many countries focus legislation in similar ways).

- The description and performance of products and services, including information about their price, are safeguarded by the Trade Descriptions Act (with a host of other laws adding weight, e.g. the Sale and Supply of Goods Act).
- The quantities of goods sold are policed by the Weights and Measures Act, so, if it says 150 grams on the packet, that is what should be there, and if goods are weighed in front of the customer then the scales had better work accurately.
- A variety of legislation affects individual industries. For example, in pharmaceuticals, food and engineering.
- Those entering into financial arrangements to buy are protected, too, by the likes of the Consumer Credit Act
- Competition, which should help provide consumers with choice and value, is maintained by the likes of the Monopolies Commission and legislation linked to restrictive trade practices.

In addition, a plethora of other influences exist. These include bodies such as the Consumers' Association (publishers of *Which?* magazine) and Citizens' Advice

Bureaux, and the fact that there is a strong European dimension to much legislation affecting consumers these days (indeed, Europe's influence grows).

Other legislation affects allied issues. The Data Protection Act means that organizations cannot hoard information about customers and use it just as they want; and other schemes, for example to prevent people receiving unwanted fax advertisements, also operate to help create the right kind of buying and selling arena. Even advertising is regulated. If it is inappropriate (e.g. racist) or simply untruthful, then advertisers can be made to stop a campaign in its tracks in circumstances that probably guarantee unwanted bad publicity, and this acts to control most excesses. Some campaigns push the limits. There are those who see French Connection's logo 'FCUK' as inappropriate, for instance. Sometimes, however, being on the borderline can be effective marketing. The industry body that polices the advertising world is the Advertising Standards Authority.

This is an area of change. Constant review leads to new or revised legislation, and new ways and styles of marketing may create new situations not covered by past regulation that need to be addressed. Opinions change, too, and that affects the attitudes of the regulators. At any point in time, however, the overall level of control is probably about right. Though some marketers might regard elements of all this as unwarranted control, most consumers approve and the climate for reasonable buying and selling remains favourable.

## 10.6   Summary

The key factor here is that understanding buyer behaviour leads to precision in the direction of promotional communication. Marketers must:

- be aware of the individuality of their customers and potential customers;
- understand how customers approach making a buying decision, and the many and various reasons that prompt them to act;
- relate this information to their market and style of product (and its potential consumer);
- act to implement promotional activity in a way that is designed to work with the reality of all this, not simply to be strident about their offering in a vacuum.

If specific promotional activity, whatever it may be – an advertisement, a brochure, a website – is predicated on this basis, then, while it still needs to be creatively executed, it stands a much better chance of having some influence than might otherwise be the case.

*Chapter 11*

# Marketing communications: the role and workings of different methods

## 11.1 What is promotion?

The word *mix* has been used in various ways in connection with marketing. The third P of the three Ps of the marketing mix is presentation (or promotion): encompassing everything about the way an organization communicates persuasively with its markets to influence them in whatever way towards making a purchase. This may be very direct: one mailshot that prompts a reply, which is an order. Or it may be much less direct. This communication happens using a variety of techniques, all rather different in their role and how they work. This chapter reviews this 'promotion mix' looking at the kinds of things that are done and what makes them effective.

This is, for many, the most interesting part of marketing; certainly it is the most visible, with elements of it – advertisements, posters and so on – all around us. It is also a very important element. As a well-known quotation (which appears right at the start of this book) says:

> They say if you build a better mousetrap than your neighbour, people are going to come running. They are like hell! It's marketing that makes the difference.
>
> – Ed Johnson

Even the producer of the best product or service will do no business, if no one knows of their existence.

Promotion is not, however, just a purveyor of information: it must be *persuasive*, and it must differentiate. Remember, potential customers may see the range of goods offered by competitors as somewhat similar. This is true of many areas, for example, cars – consider how many very similar models exist in each category – or copying machines. In publishing, the situation is certainly the same, both in areas where very similar items compete (such as dictionaries) and where distinctive titles compete with

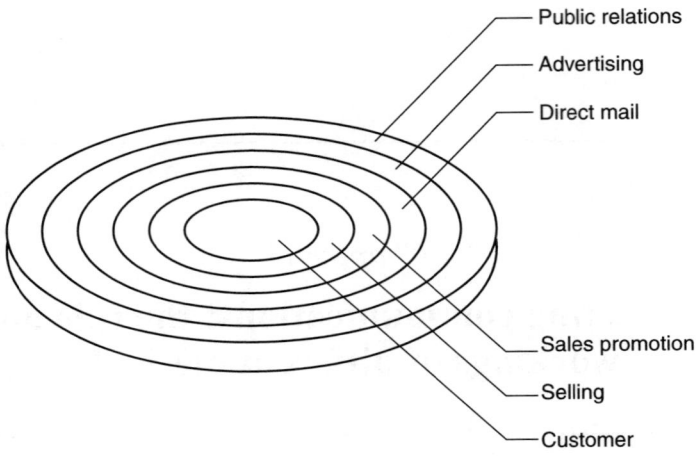

*Figure 11.1   The communications mix*

each other within a category, such as novels and gardening books. To a large extent it is the promotional elements that allow people to make a judgement about what is right for them in any product area, and this is as true of industrial items as of consumer goods.

### 11.1.1   The promotional mix

Promotion comprises a range of different elements: *mix* is the right word. In some cases the product itself acts promotionally, with the cover design of a book being important at every level of sale and display. Beyond that, the various promotional techniques themselves are not mutually exclusive: they are often used together or in various combinations – different mixes, with one or more techniques perhaps predominating. Each works in a rather different way, principally in how directly and in what way it relates to the market. The diagrammatic view, Figure 11.1, shows the different distance at which each technique operates from the potential customer; it is this that characterizes the different role which different techniques play. Notice that the customer can here be defined in any way you wish: including various people and stages along the chain (see Chapter 6).

Promotion does not act in a vacuum: it must relate to the way a potential purchaser moves progressively towards actual purchase, and act to change the attitudes of the target audience to whom it is directed. With many products it is not just purchase that is the aim, but repeat purchase, as when a householder restocks with, say, the same brand of soap. The same principle applies to books – for example, persuading:

- a college to retain a recommended book in use on a course and buy more;
- a reader to buy more of a particular author, series or category of book;

- someone to buy more books (of whatever sort) as presents;
- people to purchase sequels (which can be applied to fiction and nonfiction);
- people to favour, and trust, a publisher or imprint, prompting the way in which some people will return to the same series of, say, guidebooks.

## 11.2   The buying process

The stages in the buying process are shown in the sequence below. Each stage represents a tangible change and is a process worth examining sequentially. This is similar to points made in the previous chapter, but is now linked more closely to product examples.

### 11.2.1   *Unawareness → awareness*

This is the stage during which a buyer moves from no knowledge of a product or – in this case, any aspect of publishing (for example, an author) – towards a position where they know about it. The buyer's (which for sake of example is here equated with the reader's) attitude is nearly passive and their major need is to be informed. Promotion is targeted at:

- introducing a concept (reading business books helps you excel in your job);
- telling buyers that something, more specific, exists (Patrick Forsyth writes business books);
- creating an automatic association between the needs area and the product (if you want to find out about various aspects of business, this author is one to note).

### 11.2.2   *Awareness → interest*

This is a movement from a passive stage to an active stage of attention. The buyer will have their curiosity aroused by the product's newness, appearance, relevance or concept. Their response, however, can be conscious and/or subconscious. Promotional objectives are to:

- gain their attention through the message;
- create interest (motivation);
- provide a succinct summary of the product (information).

By this stage, aspects of the whole mix begin to get someone saying, 'I must watch out for something by this Patrick Forsyth – might be worth a look'.

### 11.2.3   *Interest → evaluation*

The buyer will consider first the effect of the product upon their personal motivation (lifestyle, image, specific needs, circumstances and so on). Then they will look at the effect on external factors around them. They will pass through a process of reasoning: analysing the arguments and looking for advantages. Depending upon their needs, they might look for more information or validation of an initial impression. Through promotion, an attempt is made to:

- create a situation that encourages the buyer to start this phase of reasoning;
- discover and focus on the buyer's relevant needs;
- segment and target buyers according to the needs requirements.

('There's a book by that Patrick Forsyth, let's have a look.')

### 11.2.4   Evaluation → trial

This is a key movement from a mental state of evaluation to a positive action of trial. The buyer's basic requirement is for a suitable opportunity to use the product. Promotional objectives now are to:

- identify clearly usage opportunities;
- suggest usage when these opportunities occur.

In other words promotion now tries to encourage first purchase. ('This does look good – and understandable. I'll buy it, just right for tomorrow's long train journey.')

### 11.2.5   Trial → usage

The buyer will take this next step if their trial has been successful. The objectives of the promotion are to:

- provide reminders of key elements, such as brand, image, advantages and so on;
- emphasize the success and satisfaction;
- remind the buyer of usage opportunities and provide supporting proof via third-party references.

Here, in our example, is the opportunity to persuade buyers to buy an author more than once. ('That last book by Patrick Forsyth was really good – I'll buy something else he's written.')

### 11.2.6   Usage → repeat usage

This is the final objective for promotion. When a buyer moves from occasional usage to constant usage, they will have moved into a state where their selection of the product is automatic. The objectives, now much simpler, are to:

- maintain the climate that has led to satisfaction;
- maintain an acceptable image;
- keep confirming the key qualities of the product.

Here an author may be promoted to encourage really regular purchase. ('These are a must – when's the next Patrick Forsyth title out? I must get it.') Although I have chosen one particular author – myself (and why not? I say – this *is* a book about marketing!), just as an example (and one published by a variety of publishers) – the point made would be valid for many authors and in different product areas too. You can fit your own organization into this frame of thinking.

The above dissects the process in detail and, though the customer may not be aware of proceeding quite so consciously through the process – and indeed may make rapid and seemingly instinctive leaps along the way (for instance adding something to their basket literally at the supermarket checkout) – this sequence is real and is what promotion must influence.

The promotional elements creating the above effect and progression can be used in any combination to make up the overall mix: everything from the effect of a small detail (such as a review quote on a book cover) to a major campaign (posters on the London Underground). With this in mind, we can turn to and review how the separate and individual techniques work, remembering that the effects they have are difficult to separate. A person's image of an organization is the net and cumulative effect of *everything* they see and hear about it.

## 11.3   Press and public relations

The term *public relations* is, perhaps confusingly, used to describe a specific technique and as an umbrella term combining public and press relations (and both are often abbreviated to PR).

### *11.3.1   Public relations*

This is concerned primarily with overall image.

#### 11.3.1.1   The impact of image

Unless it is completely invisible, every organization will have an image. But the question is whether it projects the *right* image, or sometimes whether there is a strategy to project anything at all. What do all the people to whom an organization must relate think of it? In seeking to create (and then maintain and develop) an image, it helps to have, as a starting point, a clear idea of how people see us at present. This is something that can be researched. But, of course, some information can be obtained simply by 'keeping one's ear to the ground', although, remember, people will often say what they think is expected – or, worse, what is heard only confirms the existing view, and perhaps existing prejudices.

The effect of public PR is cumulative and a host of factors, perhaps individually seeming of no major significance, are therefore important. These include the quality of business cards and letterheads, of switchboard and reception, of all printed promotion, of staff appearance – indeed, everything that contributes to overall image. A major influence also is exactly how people are dealt with, whether it is appropriate and corresponds with their image of good customer service (this is explored further in the context of sales in Chapter 12).

Consider any company that you know of, one that has a strong image and of which you think well. Then think *why* you think the way you do. Unless you have direct experience of the company, it can only be primarily because of what *it* tells you about itself. Consider too what is thought about *your* organization.

Messages projected by image can be powerful. Large companies will spend large amounts of money on their corporate image; sometimes this is obvious as activity (and perhaps press coverage) becomes unmissable. But, large or small, the image matters. Everything from the overall logo (company symbol) in all its manifestations to an individual's calling card.

### 11.3.1.2   Varying target audiences

PR must provide a planned, deliberate and sustained attempt to promote understanding between an organization and its audiences. In fact, it must promote not just under-standing, but a positive interest in the firm that whets appetites for more information, prompts enquiries, re-establishes dormant contacts and reinforces image with exist-ing customers. A sustained and systematic effort is needed and it may have various dimensions, some aimed at groups such as:

- customers and potential customers;
- government (for instance in an industry such as pharmaceuticals);
- shareholders (especially if the company is involved in the likes of a structural change or takeover);
- the city (institutional investors and similar);
- environmental and other pressure groups (to convince them ethical policies or practices are being used).

Each such group is, in the jargon, a 'stakeholder' (someone whose view or action can influence the success of the company) and constitutes a separate 'audience' towards which specific and appropriate PR activity can be directed. Some cam-paigns may be transient, some ongoing. The total list of audiences may sometimes be extensive and several different campaigns must run in tandem.

Not only is public-relations activity potentially a powerful weapon in the promo-tional armoury, it is also free – well, at least compared with advertising, which is communication in bought space. But there is a catch. It takes time! And, perhaps particularly in any small business, time is certainly money. Therefore, in too many organizations public relations is neglected because staff are busy, even overstretched, and opportunities are missed. Yet if the power of public rela-tions is consistently ignored, then at worst not only are opportunities missed, but also the image that occurs by default may actually damage business prospects. Certainly, for good or bad, many people are involved – a point Panel 11.1 makes clear.

---

### Panel 11.1   So many people affect image

Public relations links very much to a wide range of people in an organization and to standards of service. Some well-known fictitious examples occur in the award-winning training film *Who Killed the Sale?* (a film made by Rank – now Training Direct). The film shows a salesman trying, unsuccessfully, to get an order for some kind of engineering product (the product details do not matter). Progressively, we see the potential customer exposed to others in the organization and, cumulatively, such a poor image builds up that he is unwilling to place a new order – yet the salesman ends the film honestly bemused as to what went wrong.

All kinds of people help to undermine the salesman's efforts.

- A previous order is wrongly delivered, and the dispatch department, already at fault, makes matters worse by the poor way in which it attempts to sort it out.
- The switchboard operator contributes to an important message for the customer going astray when he visits the factory, as does a harassed girl in the sales office, who fails to track him down.
- A demonstration is unconvincingly conducted by a technician who fails to realize its importance because he, in turn, was poorly briefed.
- A conversation is overheard by the customer, with two more technical people being dismissive of their own company's engineering competence.
- And, the last straw, as the customer drives away from the factory he is held up, and made to reverse, by a rude delivery man who, when he closes his van door, reveals the name of the same potential supplying company.

Such a list could be wider. Try thinking for a moment about your own company. What such incidents can you remember? And can you list those people in a position to exert a positive influence on the process of building image – are you, perhaps, among them? It takes few of these kinds of incident to negate all the time and money spent on an expensive corporate image.

As we can see, in many ways time spent on PR is time well spent and, while for smaller firms it can produce good low-cost results, a larger firm able to subcontract the activity to a PR agency may well spend substantial sums. If so, they will expect to see larger-scale results, and much of that potentially comes through the press. A final thought here: it should be noted that every aspect of corporate reputations is fragile – *you are as good as your last success.*

## 11.3.2  Press relations

Press relations (a term that implies all media, rather than solely the press) is a very specific form of public relations that can pay dividends, although, unlike the case with an advertisement, there is no guarantee of what is going to be said. That said, there is no reason to feel that the media will be critical.

Though personal contact with journalists and others is important, much can be achieved through the press release, a structured, written communication to the media intended to be the basis of a story or press mention. In publishing, these lead the vital search for comment, review and news about books in the media and often, of course, a review copy of a new book has to go with them. Similarly, we read reports about, say, a new car as a result of a press function at which journalists can drive it.

Some activity is on behalf of an industry or product group, as in publishing literary awards of various sorts, such as the Man Booker Prize, which play their part in this. The opportunities here are legion: press, television, radio, newspapers and magazines of every description (some concentrating, of course, on products and organizations linked to their special area of interest).

### 11.3.2.1   The press release

Press releases can be in the form of routine mentions or of more particular stories, but remember that much of the impact of both sorts of material is cumulative. Customers will sometimes comment, 'We seem to see mentions of you lot pretty regularly,' but have difficulty remembering the exact context of what was said or, more likely, written. To achieve this cumulative impact, the PR people need to be constantly on the lookout for opportunities of gaining a mention.

Even routine matters, perhaps the appointment of a new member of staff or a move of offices, may be written up and contribute to the whole process. This is particularly so with regard to the trade press, that covering particular areas or industries. A good deal of the content there may be made up of what is often rather inconsequential news, but it helps the cumulative process of keeping a name visible. Newsworthy items must be thought of and expressed well, and are needed regularly. While some routine stories will get a mention, particularly if the company is well known, news means just what it says!

While it may be of interest internally that a firm of consulting engineers has 25 staff, inhabits an eighteenth-century mansion or is reorganizing, a journalist will tend to find it difficult to imagine readers starry-eyed with excitement as they read it in their newspaper or journal. A company must find something with more of an element of news in it: it may be genuinely different, it may be a first comment on something, but it must truly have something of real interest about it.

If a company, or its spokesperson, becomes known as a source of good comment, stories and articles, then press contacts will start to come to *them,* and the whole process may gain continuity and momentum. In addition, a number of events can be linked to press and public-relations activity. For example, both launch parties and author signing sessions fulfil this role for books, creating something more than simply the publication and launch of the book for the press to get their teeth into.

The press release that carries information of the type discussed here is a specialist document, which has to be put together just right – an example is shown in Figure 11.2 using the release to be sent out on the publication of this book. Whether this will be picked up – and quoted – currently remains to be seen, but as an example of the kind of release that may generate comment it is reasonably typical.

**The Institution of**
**Engineering and Technology**
Michael Faraday House
Six Hills Way, Stevenage
Hertfordshire SG1 2AY
United Kingdom
T +44 (0)1438 313311
F +44 (0)1438 765526
www.theiet.org

The Knowledge Network

*Demystifying Marketing: A guide to the fundamentals for engineers*
**A practical new book from the IET**

The Institution of Engineering and Technology (IET), Europe's largest professional engineering society, today announced the launch of *Demystifying Marketing: A guide to the fundamentals for engineers* by Patrick Forsyth, the first new title for the management series in 2007.

Designed to inform the non-marketing person, the book is logically arranged in a way that builds up understanding; a thorough basic guide to what marketing is, how it works and how it affects you. *Demystifying Marketing* is a comprehensive and accessible book covering all key marketing matters with an emphasis on practicality and why marketing is important in engineering.

*Demystifying Marketing* features chapters on the marketing process, pricing, policy and tactics, and marketing strategy, and aims to relate these areas and more to non-marketing jobs and experts in the engineering arena. It can provide major influence to the process of building a marketing culture in an organisation, in a way that positively influences results in the market place.

Patrick Forsyth is the author of many successful business books. One review in *Professional Marketing* magazine describes him thus: 'Patrick has a lucid and elegant style of writing which allows him to present information in a way that is organised, focused and easy to apply.'

The marketing manager at the IET said: 'This book clarifies exactly what marketing is and how it works. It demystifies the jargon and shows how marketing acts as a foundation to business success. *Demystifying Marketing* is the ideal book for anyone wanting an overview of marketing principles and how to apply them within the engineering industry, whether they are marketing novices or experienced engineers working in the sector.'

**Ends**

*Figure 11.2   An example of a press release*

So, PR is a powerful element in the promotional mix and one that most organizations utilize; indeed, as there will always be an image of some sort, by definition, then this aspect is a must.

## 11.4 Advertising

Next we consider advertising, something we all see all around us, day by day. First a definition: advertising is 'any paid form of nonpersonal communication directed at target audiences through various media in order to present and promote products, services and ideas'. More simply, it can be called 'salesmanship in print or film'. The role of advertising, as one of a number of variable elements in the communication mix, is: *to sell or assist the sale of the maximum amount of the product, for the minimum cost outlay.*

### 11.4.1  The role of advertising

Forms of advertising vary and depend on the role each needs to play among the other marketing techniques employed. There are many variations in both the type of advertising and the target to which it is directed. These include:

- national advertising;
- local advertising;
- direct-mail advertising (and leaflets inserted in journals);
- advertising to obtain leads for sales staff;
- trade advertising (to those who will resell the product);
- sector advertising (for instance, a product such as a pen is promoted to the business gifts market separately from its promotion to the retail trade).

To this list can be added a variety of electronic media these days as, for example, one business advertises on the website of another.

A more specific way of understanding what advertising can do is to summarize some of the major purposes of advertising – that is, the various and different objectives that can be achieved through using advertising in particular ways. A representative list (by no means exhaustive) of possible aims of advertising includes:

- informing potential customers of a new offering;
- increasing the frequency of purchase;
- increasing the use of a product;
- increasing the quantity purchased;
- increasing the frequency of replacement;
- lengthening any buying seasons;
- presenting a promotional programme;
- bringing a family of products together;
- turning a disadvantage into an advantage;
- attracting a new generation of customers;
- supporting or influencing a retailer, dealer, agent or intermediary;

- reducing substitution by maintaining customer loyalty;
- making known the organization behind the range of offerings (corporate-image advertising);
- stimulating enquiries (from customers and intermediaries);
- giving reasons why intermediaries should stock or promote a product;
- providing 'technical' information about something.

There are clearly many reasons behind the advertising that you see around you. Those above are not mutually exclusive, of course, and may also link to the product life cycle discussed earlier. Whatever specific objectives the use of advertising seeks to achieve, the main tasks for it are usually to:

- gain the customer's attention;
- attract customer interest;
- create desire for what is offered;
- prompt the customer to buy (either at once or in the future).

Advertising is, therefore, primarily concerned with attitudes and attitude change; creating favourable attitudes towards a product should be an important part of the advertising effort. Fundamentally, however, advertising also aims to sell, usually with the minimum of delay – although perhaps a longer time period may be needed in the case of informative or corporate (image-building) advertising. Every advertisement should relate to the product or service, its market and potential market, and as a piece of communication each can perform a variety of tasks. Thus an advertisement may:

- provide **information** – this information can act as a reminder to current users or it can inform nonusers of the product's existence;
- attempt to **persuade** – it can attempt to persuade current users to purchase again, nonusers to buy for the first time and new users to change habits or suppliers;
- create **cognitive dissonance** – this memorable piece of jargon means advertising can help to create uncertainty about the ability of current suppliers to best satisfy needs and, in this way, advertising can effectively persuade customers to try an alternative product or brand (extreme versions of this are referred to as 'knocking copy' – used sometimes by, among others, car manufacturers – and are openly critical of competition);
- create **reinforcement** – advertising can compete with competitors' advertising, which itself aims to create dissonance, to reinforce the idea that current purchases best satisfy the customer's needs – this is maintaining awareness and aiming to continue to prompt ongoing purchases.

Moreover, advertising may also act to reduce the uncertainty felt by customers immediately following an important and valuable purchase, when they are debating whether or not they have made the correct choice. This is perhaps just as important in most fields of engineering as in any other area, but it is all part of constant reinforcement.

Clarity is always important. If an advertiser is, say, a charity, trying to raise funds to help starving children, the purpose of its advertising and what it hopes 'customers' will do will be clear to everyone: send a cheque. Sometimes the action may be less clear and, at worst, confusion can reign and no action succeed in being prompted. As an example of how creativity can produce something clear yet punchy, I cherish the sign that appeared in an outdoor-wear shop in Stratford-upon-Avon announcing its January sale. It said, 'Now is the discount of our winter tents'. And it's memorable, too.

## 11.4.2   Types of advertising

There are several basic types of advertising and these can be distinguished as follows.

- **Primary:** This aims to stimulate basic demand for a particular product type: for example, insurance, tea or wool.
- **Selective:** This aims to promote an individual brand name, such as a brand of toilet soap or washing powder, which may be promoted without particular reference to the manufacturer's identity.
- **Product:** This aims to promote a product or range of related brands where some account must be taken of the image and interrelationship of all products in the mix.
- **Institutional:** This covers PR-type advertising, which, in very general terms, aims to promote the company name, corporate image and the company services.

And, of course, there is a variety of different media.

## 11.4.3   Advertising media and methods

There is a bewildering array of advertising media available. Here are some of the most popular methods of advertising, with a guide as to how they are used. All are potentially appropriate, but a mix must be selected that fits the purpose of organization and product. The main categories, which tend to provide good awareness among consumers, include the following.

- **Daily newspapers** often enjoy reader loyalty and, hence, high credibility. Consequently, they are particularly useful for prestige and reminder advertising. As many people read them hurriedly, lengthy copy may be wasted.
- **Sunday newspapers** are read at a more leisurely pace and consequently greater detail can be included.
- **Colour supplements** (and similar) are ideal for general advertising, but appeal to a relatively limited audience, and tend to be expensive.
- **Magazines** vary from quarterlies to weeklies and from very general, wide-coverage journals to many with a specific focus and some linked to very specialized interests. Similarly, different magazines of the same type (such as women's magazines) appeal to different age and socioeconomic groups. Magazines are normally colourful and often read on a regular basis.
- **Local newspapers** are (obviously) particularly useful for anything local, but are relatively expensive if used for a national or broader campaign. They are sometimes used for test-market-area advertising support prior to a national launch.

- **Television** is regarded as the best overall medium for achieving mass impact and creating an immediate or quick sales response. It is arguable whether or not the audience is captive or receptive; but the fact that television is being used is often sufficient in itself to generate trade support. Television allows the product to be shown or demonstrated and is useful in test-marketing new consumer products because of its regional nature, but is very expensive, and therefore ruled out for almost anything except mass market products.
- **Outdoor advertising** lacks many of the attributes of press and television, but it is useful for reminder copy and a support role in a campaign. Strategically placed posters near to busy thoroughfares or at commuter stations can offer very effective, long-life support advertising. Collaboration between manufacturer and retailers can link these to strategic locations designed to support local activity.
- **Exhibitions** generate high impact at the time of the event but, except for very specialized ones, their coverage of the potential market is low. They can, however, perform a useful long-term prestige role. In specific ways some can be very effective and, like many industries, the engineering world has ones that those within their field regard as unmissable. (See more on exhibitions below.)
- **Cinema**, with its escapist atmosphere, can have an enormous impact on its audience of predominantly young people; but without repetition (people visiting the cinema once every week, or a tie-in with other media) it has little lasting effect. It is again useful for backing press and television, but for certain products only, bearing in mind the audience and the atmosphere. This is another medium where cost is high.
- **Commercial radio**, playing music for every conceivable taste, or focusing on interest groups (news or phone-in freaks, or whatever) offers repetitive contact, has proved an excellent outlet for certain products, and is expanding its users all the time. It is becoming apparent that the many local radio stations appeal to a wide cross section of people and thus offer support potential to a wide range of products.

Not all these (or direct mail, reviewed later) are right for every product or in every circumstance. Some are simply not cost-effective in every circumstance. Others assume greater importance because of their other linked characteristics – an advertisement alongside a good press mention may work much better than one without this editorial link.

Another technique worth further mention here, one used in some areas of engineering, yet not really advertising, is exhibitions. Exhibitions (and trade fairs and the like) constitute an area that overlaps with that of selling. Advertising needs to attract people to the exhibition; the design of the stand needs to create appropriate impact once they are inside; and the people on duty need to play their part well. For example, nothing switches visitors to an exhibition off more quickly than the stereotyped 'Can I help you?'. To which the reply is usually, 'No, thank you, I'm just looking'. Not only does the personal input on the day have to be good, but so too does the follow-up contact, which must be prompt and appropriate. Exhibitions are hard work – as well as hard on the feet and the stomach! – but they can be very useful in generating business opportunities.

Every advertiser must make its own decisions (advertising agencies that handle the larger advertising budgets have sophisticated media-buying departments), decisions not only about different methods, but also about exact media – one newspaper versus another and so on.

Not all advertising, however, is aimed at potential consumers: some is directed at intermediaries.


### 11.4.4   Trade advertising

This is certainly important in many industries. It is often not sufficient to advertise products to consumers alone, particularly where it is important that distributors/retailers are willing to stock and promote a product.

Even though the sales force often has a prime role to play in ensuring that stocking and promotional objectives are achieved, trade advertising also has an important role to play in this respect. Indeed, trade advertising sets the scene for such sales visits, since it can:

- remind intermediaries about the product between sales visits;
- keep them fully informed and up to date on developments and changes of policy;
- alleviate problems associated with the cold-call selling of lesser-known products;
- indicate the support, and weight, being given to a product – this is disproportionately important within the trade, being used both as an objective measure of assessing what stocking to decide upon, and as an easy (albeit subjective) criterion in making a quick decision.

Trade advertising defines communication with specialist trade publications (advertising and direct mail); often trade advertising occurs prior to, or linked to, consumer advertising campaigns to help prompt the buying in of stock in anticipation of future demands to be created by the consumer advertising. Other tactics link in. For example, when new products are launched, or special promotions introduced, trade support may be achieved through special offers (an incentive if you order a minimum quantity), or increased (introductory) discounts, all of which trade advertising can effectively emphasize.

This type of advertising can also communicate to the trade the detail of the moment – why they should stock forthcoming new products or the existing range, as well as flagging the timing and weight of any advertising or other promotional support that is to come.

Curiously perhaps, one of the most important aspects of trade advertising is not what it says (though I do not intend to suggest it does not matter what is said or how it is put over, for it does – and we will cover more of this later). It is *the fact that it is there*. The commitment (and cost) of taking such space is seen as a commitment to particular products. If a salesperson is pressing someone to take good stocks, instigate their own promotion or generally take a product seriously, then buyers are apt to ask,

'What are you doing for it?'. Reasonable enough: it will be easier to sell something if the supplier is taking action to make people want to buy it.

Even so, the advertising mix that is deployed here, whether it is directed to the company, product range or brand, must accommodate this fact. The more that retailers feel that suppliers are matching their views of potential success with action, the more likely they are to respond positively to stocking suggestions.

Advertising effort needs to be spread among the various target audiences that match a particular product. Trade advertising may take a large share of this in some industries. It must be tailored to the trade, which probably wants to hear different things from the ultimate consumer, and be spoken to in a different way. The basic principles of what makes advertising work and the strategies involved are similar for all types of advertising. And it is to these issues we move on now.

## 11.4.5  Advertising strategy

Advertising needs to be designed and produced in a way that reflects an analysis of the market and a subsequent sensible choice of media and advertising strategies. This means that those involved – more than one person is often involved – must be in close communication. It may help to spell out what needs to be done in a simple strategy document. At its best, such an advertisement strategy statement is brief and economical, and does its job in three paragraphs describing:

1. *the basic proposition* – the promise to the customer and the statement of benefit;
2. *the 'reason why'* – the support proof justifying the proposition, the main purpose of which is to render the message as convincing as possible;
3. *the 'tone of voice'* – the manner in which the message should be delivered, the image to be projected, and not infrequently the picture the customer has of themselves, which it could be unwise to disturb or, rather, wise to capitalize on.

In various fields some of the finest and most effective advertising has sometimes been produced without reference to any advertising strategy, or for that matter without knowledge of real market facts – art and science again. However, although research or objective thinking cannot always give all the details, or for that matter always be infallibly interpreted, it can give strong indications and reduce the chances of failure.

Most executives, when faced with a rough or initial visual and copy layout, have an automatic subjective response: 'I like it'/'I don't like it'. And, while the creator may attempt to explain that the appraiser is not a member of the target audience, it can be genuinely difficult to be objective. Nevertheless, while an attempt at objectivity must be made, there are few experienced advertising or marketing executives who can say that their judgement has never let them down. The most famous saying about advertising is that of the company chairman who said, 'I know half the money I spend on advertising is wasted, but I don't know *which* half' – a remark that contains a good deal of truth, and a sobering thought in any organization where every penny of the budget has to be fought for.

Another possible problem in any business is that of 'me-tooism'. Advertising gets into a rut; those producing it simply reiterate an established formula and cease even to try to think creatively. This gives rise to so-called tombstone-style advertisements consisting of little information or inspiration. Easy to produce, low in cost, yes – but hardly likely to have major or striking impact. Such an ad might be judged fine as an announcement to the faithful (for example, those who will buy a well-known author's new book automatically), but surely much advertising is, or should be, designed to do more than that.

It is one thing for it to be there, visible; it is quite another for it to be *persuasive*. There is a danger also of confusing creativity – the process that makes something both appropriate to customers and memorable – with cleverness. Sometimes a clever idea – a play on words in a headline, perhaps – can act not to increase the power of the advertisement, but to dilute it or obscure what should be a clear message. Advertising must never fall into the trap of confusing cleverness with clarity of communication.

All companies must ask straight questions about their advertising generally and also about particular advertisements. Specifically they should ask the following.

- Does the advertisement match the strategy laid down?
- Does the advertisement gain attention and create awareness?
- Is it likely to create interest and understanding of the advantages of what it offers?
- Does it create a desire for the benefits and really prompt the need to buy?
- Is it likely to prompt potential customers actually to make a purchase, now or in the future?
- Can it be linked to tangible action (with a coupon to be completed and returned, a 'hotline' to be telephoned, for instance)?
- Is it concentrating on the features of the product, rather than benefits to the purchaser?

In other words: does the advertising communicate? Will people notice it, understand it, believe it, remember it and buy as a result of it?

The next question is: how can an advertisement be made creative? There are many ways: humour, personalities, exotic locations, cartoons, even running advertisements in the form of a serial, ending with a cliffhanger to encourage viewing the next. Gold Blend coffee was the first to use this, albeit on television, at least in the UK, and it even ran press ads saying no more than the time and date of the next instalment. It is an example that illustrates that new ideas can be found, and sales can rise as a result.

In other words, advertising needs to be *creative*. Often its task is to make something routine, or even potentially dull, seem *interestingly different*. Just occasionally the product really is interestingly different; more often the essential qualities of the product need presenting in whatever way allows the presentation to persuade. There are many ways of doing this and the example that follows illustrates this further.

### 11.4.6   Advertising approaches

The following invented illustration reviews the options in a way that makes clear the approaches involved, albeit doing so in a light-hearted way. Sometimes the product is such that with no competition, with a perfect match with customers' needs, all the advertisement has to do is say what the product will do for them, for example, 'New instant petrol – one spoonful of our additive to one gallon of water produces petrol at 1p a gallon'. If your message is like this, there is no problem; persuasion is inherent in the message. But few products are like that.

More likely, any product will have competition of one sort or another. Then you have to say more about it. Start by thinking of everything about it. You may even *say* everything about it:

> SPLODGE – the big, wholesome, tasty, non-fattening, instant, easily prepared, chocolate pudding for the whole family

Or you may stress one factor, thus implying that your competitors' products are lacking in this respect:

> SPLODGE – the easily prepared pudding

Customers may know all puddings of this sort are easy to prepare, but they are still likely to conclude yours are easiest. The trouble with this approach is that, in a crowded market, there are probably puddings being advertised already as 'easily prepared' – and big and wholesome and all the rest for that matter. What then? Well, one way out is to pick another factor ignored by your competitors because it is not essential:

> SPLODGE – the pudding in the ring-pull pack

It may be a marginal factor but your advertisement now implies it is important and that competition is lacking. Alternatively, you can pick a characteristic of total irrelevance:

> SPLODGE – the pudding that floats in water

Or link it to the pictorial side of the advertisement:

> SPLODGE – the pudding you can eat on the top of a bus

If the competition has done all of this, then you have only one option: you must feature in the advertisement something else, nothing to do with the product. This may even necessitate giving something away: 'SPLODGE – the only pudding sold with a *free* sink plunger'; or repackaging it: 'SPLODGE – the only pudding in the *transparent* ring-pull pack'.

The possibilities are endless and the ultimate goal is always to make your product appear different and attractive, and desirable because of it.

In addition to all this, advertising has to be made to *look* attractive. Sometimes this may be achieved through the added humour, personalities or whatever, or through lavish production values – just commissioning special photography may be costly, but the quality may immediately create something special. A danger here is that the pluses hide the message: viewers of a poster, say, laugh at its humour but cannot recall exactly what was being advertised on it.

In increasingly competitive times more creative approaches to advertising are always necessary and, when found, can pay dividends in terms of results.

*Table 11.1    Attention, interest, desire, action*

| Objectives | Methods |
|---|---|
| Gain customer's **attention** | Select right media; buy enough space; impact headlines/pictures |
| **Interest** reader | Involve customer; make big promises; solve problem; present facts; communicate quality |
| Make reader **desire** benefits of your service | Show benefits for them; don't bore – enthuse; use testimonials; message clear and simple; be honest |
| Get **action** from reader – make a sale | Summarize benefits; ask reader to buy, call, telephone, attend an event, fill in coupon, make a choice, save time and money, buy now; ask for order |

All sorts of 'formulae' exist to focus thinking about advertising and ensure that impossibility, the perfect advertisement. Table 11.1 is typical of such approaches and certainly here acts to illustrate the thinking involved.

## 11.5   Direct-mail promotion

We turn now to a very particular form of advertising. Although direct mail is just, in effect, a different form of advertising, it is sufficiently important to merit a few words under its own heading and allows more to be said about the copy (the actual message to customers). It is an important, useful and effective medium. Only the worst of it should be called 'junk mail', though some perhaps deserves that description – as Samuel Johnson said, 'What is written without effort, is in general read without pleasure'.

It can be directed at any level of the buying process, from ultimate customers to specialist intermediaries (it is sometimes useful for customers who do not justify the cost of visits, or more than a certain number of visits). It is used as much in business-to-business activity as in communicating to consumers in the traditional sense.

Somehow feelings about direct mail seem to run high. Some people regard it as intrusive. Everyone appears to know someone who has been mailed three times in the same week about something entirely inappropriate, and addressed wrongly as 'Dear Madam'. Some people regard it as more than intrusive, ranking it somewhere between picking your teeth in public and being unkind to animals.

Direct mail is, though, only a form of advertising – no more, no less – albeit a specialized form. It is used very successfully in a wide range of industries and applications, many of them perfectly respectable – charities, banks, building societies and so on. Many others are dependent on it as their main form of promotion or because it results in a major proportion of their sales (indeed, one example of this is business books). What is more, although of course there is the occasional annoyance, it is used for the most part without upsetting the people to whom it is directed. Indeed some will even pay to receive it. I pay to receive mailings from the Barbican in London, and another example is a 'catalogue' such as *The Good Book Guide*.

If people are not interested, they throw direct-mail material away, a process that is not really so unlike turning over an advertisement page in a magazine in which one is not interested. Of course direct mail is wasteful. It hurts to think of so many carefully penned words ending up in the bin (at least, it hurts the originator!). But it is no more wasteful than other forms of advertising. All advertising is in a sense wasteful – what matters is whether it produces a cost-effective response, whether it pays for itself in the long term.

Contrary to popular belief, direct mail is read. In the UK, the Post Office, which spends a great deal of time and money studying the effectiveness of direct mail, recently demonstrated through research that more than 90 per cent of it is opened and more than 75 per cent of it is read. The trick is less to achieve this, therefore, than to ensure your offering will stand out from others, will generate interest and will be seen as persuasive.

Direct mail is not an alternative to advertising: rather, it adds to the range of techniques available. It is no more a magic formula than any other individual technique. But it can sometimes suit well. It is flexible, certainly more flexible than advertising. Direct mail may constitute four letters, or 40, 400, 4 000 or 40 000. It does not *have* to be done on the grand scale: it can be targeted at small specific groups or it can be undertaken progressively, with so many shots per week or month being sent. It is personal and can be directed at specific and discrete groups.

It is controllable, it can be tested, implemented progressively and results can be monitored to ensure it provides a cost-effective element in the total promotional mix. Since it is likely to be low-cost per contract, and campaigns can be varied so much in size, there are likely to be many more organizations that could experiment with the technique. It can be specific or it may be directed broadly, selling the firm, or be part of the promotion of particular products.

So, direct mail has a lot going for it. It is a proven technique in many fields. It can be used on a small scale; it can be targeted at specific market segments; and it can be tested and monitored much more easily than many other forms of promotion. But it is also deceptive and can appear easier to use than, in fact, it is. Every element of it needs careful consideration.

## 11.5.1   The elements of direct mail

Briefly, each element of direct mail contributes to its success. Such include the following.

- **The list:** Any mailing is only as good as the list of names it is mailed to. It must be appropriate, up to date and personal. Mailings addressed to an individual do best. In many businesses existing customers are as important as prospects, so complex overlapping campaigns are constructed; there is a specialized area of 'database marketing' and, though list holding and use are now covered by the Data Protection Act in the UK, sources of lists are valuable. Next time you are asked in a store to write your name and address on the back of the cheque, that may be one of the reasons. Other techniques, such as satisfaction cards, are specifically designed as list builders.

- **The message:** This is vital. Copywriting in this area is a specialist job. Just one phrase changed may increase (or decrease) the response. There are about three seconds, when something is pulled out of an envelope, during which time the recipient decides whether or not to read on further, so immediate impact is crucial.
- **The envelope:** The 'packaging' is part of the message – many envelopes are overprinted, perhaps with a 'teaser' message. What is on them affects response. It is particularly important in assisting the job of getting the recipient to decide to read on.
- **The letter:** This is also vital and is often not short. A good message is as long as is necessary to present an argument to buy, and, if this takes two or three pages, so be it; many letters are longer and work well. As a general rule, a brochure of some sort plus a letter pulls better than a brochure on its own.
- **Brochures:** These provide supporting information in a profusion of ways. They may be coloured, illustrated or, in extreme cases, incorporate a range of gimmicks such as prize draws.

Direct mail is a technique where tiny details matter. For instance, a letter with a PS may do better than one without; a reply card with an actual postage stamp (rather than with prepaid postage) may get up to 50 per cent more replies; and certain so-called 'magic' words (*new, free, guaranteed, exciting*) seem to boost response, provided they are not overused.

Be aware that the copy – all the text, that is, whether in letter or brochure – may not be read. People dip in and out of the material, so some repetition of content between, say, letters and brochures (though perhaps not actual words) may be sensible. Certainly the relationship between what is said in different elements of the shot needs careful consideration.

All this certainly gives the feel of what is necessary to implement direct mail successfully on any scale. It links to planning and research; and is not possible without clarity of purpose and a sound base of information. Clearly, it does not just happen. The kind of thinking demonstrated here is important (and valid not only in terms of direct mail, since a similarly thorough approach would strengthen any promotional campaign). But the potential is clear, too. With this in mind, you may like to look more carefully at the next shot you receive; it may encapsulate a great deal about how marketing works.

As we have seen, the success of direct mail is very subject to the *detail* of what is done; something as simple as a different headline on literature or letter may sometimes significantly affect response. But it has many virtues as a promotional method – not least its ability to be tested (a small shot may be sent to judge its effectiveness before more money is spent circulating it more widely).

It should be noted that many of the principles of direct mail apply also to newsletters and other material received by email. Much of this is directed at existing contacts; it is either requested or appreciated and found useful by recipients and works well.

Unsolicited email – referred to as spam – is a different matter. A modern plague, it only annoys but its use is thus likely to be self-defeating and it is not recommended. Hopefully, legislation and technology will eventually reduce or mitigate this.

## 11.6 Sales promotion

In formal marketing terms, sales promotion can be defined as 'an inducement aimed directly at persuading a specified target audience to achieve one or more defined objectives'. In simpler terms, it is a method of persuading people to take a course of action that, without that persuasion, they would not otherwise take. It encompasses areas such as point-of-sale (POS), offers, giveaways, and merchandising and display in retail situations.

Sales promotion is an aid to selling, not a substitute for it. It is a tactic that is used because, after careful analysis of the facts and quantification of the objectives, it is likely to prove the most cost-effective method of meeting those objectives.

Sales promotion is not the answer if:

- all else looks doomed to failure;
- the sales force has a period of inactivity;
- someone thinks it would be nice to have some;
- the chairman's wife had a good idea.

Yet, perhaps because it is an 'ideas' area, this can be precisely what happens. If the company tries to make a problem fit an idea, instead of creating a planned scheme to solve the problem, then experience shows that there is a very great chance that it will not work. Remember that, as with any area of promotion, if it is not precisely planned and controlled, it could well have the directly opposite effect to the one it set out to achieve.

### 11.6.1 The role of sales promotion in marketing

Sales promotion is not an incidental technique that should be used only as a last-minute afterthought; nor should it be left to junior staff to plan and implement. It is an integral part of the marketing mix, and, as such, requires the same degree of planning, as does the mounting of a market-research project, or the selling-in of a new product. Effective planning is therefore essential, whether sales promotion is to be used as a support activity for the company's long-range objectives or as a short-term tactic.

A more specific way of understanding what sales promotion can do for the company is to review some of the major purposes of sales promotion, or the objectives that can be achieved through using it effectively. For example, it can be used to:

- **introduce new products**, by motivating customers to try a new product or intermediaries to accept it for resale (for example, introducing the idea of two products linked to sell together);
- **attract new customers**, by motivating existing customers to try a new product or retail customers to accept it for resale (for example, so someone buys a particular product because of a competition they can enter);

- **maintain competitiveness**, by providing preferential discounts or special low prices to enable more competitive resale prices to be offered;
- **increase sales in off-peak seasons**, by encouraging consumption 'out of season';
- **increase trade stocks**, by special monetary discounts or quantity purchasing allowances, in return for holding greater than normal levels of stock;
- **induce present customers to buy more**, by competitions to encourage customers to think of more ways and more occasions for using the product.

Generally, then, sales promotion is a marketing device to stimulate or re-stimulate demand for a product during a particular period. It cannot overcome deficiencies in a product's style, quality, packaging, design or function, but can provide an important addition to advertising activities as an integral part of the communications mix.

## 11.6.2    Types of promotion

We will now turn to some examples of what can be done. The list is no doubt not comprehensive, as a new promotional idea is thought up somewhere every minute of the day. Indeed, there are no hard-and-fast rules for selecting the 'right' sales promotion tactic, since what is successful in achieving an objective in one situation may not necessarily be successful in achieving a similar objective in another. Similarly, the same promotion tactic might be suitable for meeting different objectives.

In practice, there are likely to be many options, all of which would be suitable for meeting the same objective. Selection can be assisted by asking (and answering) questions such as the following.

- Which promotion tactic best fits the profiles of the target audience?
- What are the advantages of each promotion tactic?
- What are the disadvantages of each promotion tactic?
- Which is likely to give the greatest level of success for the budget available?
- Which promotion best lends itself to accurate measurement of its effectiveness?

The types of promotional tactic currently available are many and, while they cannot be strictly confined into set categories, those discussed below show something of the range of what goes on using a number of simple everyday examples.

### 11.6.2.1    Promotions received at home (or in an office/factory)

In-home consumer promotions can help to pre-empt the attempts of competitors to solicit impulse purchases via in-store advertising and display. Techniques used here include:

- sampling, where a sample of the product is delivered free to consumers' homes;
- coupon/voucher offers via postal and door-to-door distribution, newspaper or magazine distribution, and in-pack/on-pack distribution;
- competitions.

### 11.6.2.2 In-store promotions

Clearly this type of promotion has the major advantage in that it is featured at the location where many of the *final* decisions and actual purchases are made. Techniques used here include:

- temporary price reductions;
- extra-value offers, including those relating to future purchase (this includes store cards that collect points as with the likes of Tesco);
- premium offers (incentives), including free mail-in premiums, self-liquidating premiums and banded free gifts;
- point-of-sale product demonstrations;
- personality promotions (with a demonstrator – as in toy shops where people show off things such as robot toys – or with someone famous such as novelist signing their books).

### 11.6.2.3 Immediate-benefit promotions

Here, consumer reward for purchasing is immediate, and, as with most incentives, the sooner the reward can be expected and received after the qualifying action, the greater will be the positive effects of that incentive in stimulating purchase action. Included in this promotion category are:

- price reductions;
- free gifts (which can be additional products, such as two for the price of one; this is now in the language as BOGOFs – buy one, get one free);
- banded pack offers (for instance a new razor with shaving soap);
- economy (special editions and own-brand items).

Some of these are offered by retailers as part of other broader schemes. For example, you can obtain Shell Smartcard points (or Airmiles) at John Menzies and other retail outlets, as well as with your petrol.

### 11.6.2.4 Trade promotions

Some promotions are directed exclusively at intermediaries or their staff. The reasons for promoting to the trade include:

- obtaining support and cooperation in stocking and promoting products to customers;
- inducing distributors to increase their stock levels, where research may have revealed lower than average stockholding;
- pre-empting competitive selling activities by increasing trade stocks.

Among the techniques used in trade promotion are:

- **Bonusing:** This can take the form of monetary discounts or 'free goods' (13 products for the price of 12), or special quantity-rate terms.
- **Incentive schemes:** These can be tailored to the needs of a retailer's sales staff and may also include competitions, particularly for sales staff. Competitions linked

to generating window displays make a good example from retailing, with prizes such as holidays being regularly used.

- **Dealer loaders:** Instead of money, gift incentives may be offered to distributors, or their sales force, for achieving agreed sales targets or stocking certain quantities of a product.

Other trade promotions are linked to publicity rather than directly to sales.

- **Cooperative advertising schemes:** These give assistance with preparation of advertisements or media costs, which in some businesses make promotional activity possible that might not happen unsupported.
- **Provision of display materials:** These can be either free of charge or on a shared-cost basis; includes things from display stands to window stickers and shelf talkers.
- **Tailor-made promotions:** These are custom-designed to the outlet's individual requirements, often promoting their own name and corporate image.

Thus, it can be seen that trade promotion can be an extremely important element within the total market strategy in helping to ensure that stocks are available in the right distribution channels and at the right time – and are actively promoted at point of sale.

### 11.6.3　Sales promotion in action

While sales promotion was pioneered in the area of FMCG (fast-moving consumer goods) products, where it is most visible, it is also used – albeit in slightly different ways – to influence any kind of customer. Thus sales promotion might be needed to:

- encourage repeat purchase;
- secure marginal buyers;
- meet competition;
- ensure that bills are paid on time (or sooner!);
- motivate a retailer's sales assistants;
- induce rapid market penetration when launching anything new;
- sustain perception of value over and above that intrinsically possessed by the product itself;
- smooth out costly buying cycles and seasonality.

If buying behaviour can be changed in these ways, to whatever extent, then the supplier will be more productive and profitable. Promotions are not always effective, but when they are they may have a considerable effect; and to achieve that they need to work rather than be unique.

## 11.7　Merchandising and display

Usually regarded as being under the promotion umbrella but deserving a heading of its own is something else applicable to retail marketing: that is the area of merchandising and display. This is the term given to the promotional effect of layout and display in retailers of all sorts, though elements of it may be just as important in somewhere like an industrial showroom. This is a section you might skip, though as a shopper you may find that it will explain why some things are as they are.

Have a look around you next time you go shopping. Are there shops with window displays that make you want to look inside? Do you notice, in a supermarket, that essentials, such as bread and sugar, are very often at the back of the store? This necessitates customers passing many other, less essential items *en route* to those they really want to buy.

The phrase 'impulse buy' is used to describe purchases made on the spur of the moment because something catches the eye. This is not a way of making people buy something they do not want, so much as a way of making sure they buy sooner, rather than later and from somewhere else. So things that need some promoting tend to go at the front of the shop, and staples (bread in a supermarket) to the back.

Merchandising and display have clear objectives, and research confirms they are a key influence on purchasing decisions. Merchandising and display are designed to achieve the following.

- **Sell more** – that is, to sell a quantity over and above the level that would occur if no action were taken. Some people will always want certain products and will search them out.
- **Inform the customer** about numbers of matters in numbers of ways: telling them a shop is there, indicating something of the range of products it sells, highlighting what is new, directing people to the right section of the shop, and so on.
- **Persuade**, making the message attractive, understandable and convincing – it is this aspect that can prompt the action that is really wanted: a sale!

They have to put over messages to many different groups of people, perhaps particularly three:

- those who might otherwise pass the shop by, who will not even enter unless something external catches their eye;
- those who come into the shop for one small item and who may buy more, and the ubiquitous 'browser'; so shops that attract browsers stock a range of different things (as with bookshops and greeting cards);
- those who are active or regular customers.

Of course, there are all sorts of people within each category, young, old, richer, poorer, male, female, and so on. Because of their different intentions some messages will be general; others will be specific, aimed exclusively at one group or another. In addition, there are the products themselves. To say the promotional and display permutations become numerous is an understatement. In many shops – certainly bookshops – the number of lines stocked is numbered in thousands.

Any change of products to be sold (and therefore displayed), coupled with the customers' tendency to notice only what is new, means displays must frequently be changed or updated. Display is there, in part, to remind and to freshen the interest. Again, if you consider your own high street, you may notice, let us say, a window display instantly if it is eye-catching (you may even go into the shop and buy something). However, if the window is never changed, it just becomes part of the scenery and, after a while, it makes no impression.

## 11.7.1   Fascinating AIDA – how to do it

Display, therefore, must be carefully carried out to achieve the right effect. There is a mnemonic, AIDA, which demonstrates exactly what needs to be done:

A catch the customers' **attention**
I arouse their **interest**
D turn their interest into **desire**
A prompt **action**

This mnemonic can be applied to many promotional intentions. Thus a customer seeing a display of books labelled 'for holiday reading', to take a simple example, has their eye caught by perhaps one aspect of the display – a bucket and spade – and starts thinking, 'What's this?'; 'Perhaps I do need a book for my holiday'; 'That looks just the thing' (looking at something specific); 'I'll buy it'.

This is the essential principle behind good display, and checking a particular display to see if it will carry customers through this kind of sequence is a useful test of its likely effectiveness. Another thing here is ringing the changes and surprising people. All sorts of interesting combinations are no doubt possible.

## 11.7.2   The physical layout of the shop

It is beyond the scope of this book to consider the physical layout of a shop in detail. In any case, many aspects are fixed, for a variety of reasons: cost, the lease will not allow change, and so on. Others are not, and certain basic principles of layout are worth commenting on briefly, as they are certainly part of the promotional mix. So, in no particular order of importance, consider the following.

- **Traffic flow:** Ninety per cent of the population are right-handed and will turn left on entering a shop and tend to go round it clockwise. (This is compounded by habit as so many supermarkets and department stores, recognizing this, encourage it – it has then become the norm with many of us.)
- **Eyes:** Customers select most readily from goods set out at eye level (60–62 inches/1.52–1.57 metres for a woman, a little higher for a man). This puts very high or low shelves at a corresponding disadvantage – and many shops have plenty of both. There are problems here with the volume of stock to be carried and displayed, but customers may resent having to shop on their hands and knees.
- **Quantity:** Customers buy more readily from things displayed in quantity, rather than a single example of a product.
- **Vertical display:** Products displayed together are found more manageable if they are above and below one another rather than arranged side by side.
- **Accident:** Customers are less likely to pick up or browse from any layout that appears accident-prone – that is, if they think they may not be able to balance an

item back in position or that other items may fall, especially if they fear damaging something. And this is important in any shop where the customer needs to pick up and inspect products.

- **Choice:** Customers are attuned to choice. A number of options make this easy to exercise – products sell better from within a range of similar items.
- **Relationships:** Customers expect to find related items close at hand (so, gin and tonic, strawberries and cream – clever combinations may help).
- **Cash points (tills):** These need to be convenient and clearly indicated (and, of course, promptly and helpfully staffed) and can be a focal point for some display.
- **Position:** In a large shop, people will walk or search further for things they feel are essential (as we have seen, it is no accident that bread is normally at the rear of the supermarket). So, if the children's clothes are up three flights of stairs, mothers with pushchairs may not make that shop first choice.
- **Colour:** This has a fashion, and an image, connotation – bright may be seen as brash – so it must be carefully chosen. This applies to display materials, such as a backcloth in the window, as well as decoration. Too dull, however, and it is not noticed.
- **Lighting:** This must be good – perhaps especially in a bookshop – because if something cannot be found or seen clearly no one will buy it, and people's patience is limited.
- **Seating:** Some shops want to encourage browsing, so, if lack of space does not prohibit, they provide some chairs, and perhaps a stool near the till for older customers.
- **Background music:** This evokes strong opinions. Some like it; some hate it. However, the reverse, a library-like silence, can be offputting for some. Certainly careful choice and consideration of volume level are necessary. Some shops even display a notice to tell you what is playing!
- **Character:** Part of the overall atmosphere will come from the main physical elements of the shop – dark wooden panelling has a quite different feel from more modern alternatives (both may have their place).
- **Floor:** This will be noticed. Is it quiet? Can it be kept clean easily? And does it (or should it) direct customer flow as some shops do, using different colours for pathways?
- **Reach:** If things are out of reach, people are reluctant to 'be a nuisance' by asking and may not buy.
- **Signs:** Again, since people are reluctant to ask, there must be sufficient signs, and all must be clear and direct people easily to everything in the shop. In addition, many signs are virtually in-store advertisements, and these can be used to good effect, perhaps especially if they are striking or unusual.
- **Standing space:** Space to stand and look without completely blocking 'traffic' flow will encourage purchase.
- **Security:** Last, but by no means least, this is sadly important in every aspect of retailing, from the equipment, such as closed-circuit television that may be used, to simple vigilance. Retailing is unfortunately rarely sufficiently profitable to sustain a high level of pilferage without concern.

This list is not, of course, comprehensive, but these and other factors are important and the overall physical construction and layout are the backcloth to any display, in the shop or in the window. Merchandising perhaps demonstrates the need to leave no 'promotional corner' ineffective.

Finally, this is an area that can demonstrate the bizarre end of customer behaviour. For instance, both smell and music trigger sales. A supermarket will sell more fresh bread if the smell of it pervades the store and sell more French wine if the music playing within sight of it has a French sound. Why this happens is best explained by the biologists and psychologists, but if this sort of thing works, and it does, marketers will use it.

## 11.8   Electronic promotion

At the risk of making this a long chapter, a number of things are dealt with here all in context of what might be called the electronic revolution. Note from the start that it is the *way* of doing things that is changing. A website essentially, for instance, fulfils the same role as a brochure and, if you bought this book via Amazon, you are only utilizing one marketing channel rather than another, using a website that is both a brochure and a distributive method. But things are changing. Let's start by looking at how much and in what way, before homing in on the promotional aspects.

### 11.8.1   The Electronic Revolution

This section is out of date. Even as I write it, the pace of change means that, by the time it has got into print and you read it, things will have moved on. That said, it should not matter, at least within a reasonable time frame, since what is most important here are the trends and the attitude to information technology that must be taken with regard to marketing.

Some general points first. The pace of change is frantic. Even considering the longer term, things move fast. Computers have revolutionized the office environment, but not everything has progressed as predicted. Consider the following.

- Computers themselves have perhaps evolved faster than initial predictions, but what, for instance, has happened to the 'paperless office'? Most people's desks seem as submerged as ever (and some of what covers them is computer printout!).
- Computers make things more efficient and do things faster, but how long does it take to get to grips with the latest feature, and why have such phrases as 'sorry, it's in the computer' become the ultimate excuse for delay?
- Computers have reduced costs, but what about the cost of the equipment and the training and the peripherals?

These and no doubt other statements that could be made and questioned expose two sides of the proverbial coin. There is truth in both aspects of them. It is too simplistic to expect just to be able to say, 'Computers make things better' and expect there to be no downsides. The point about change remains. It took some years for computers to become established in every office, proportionately less for them to proliferate onto every desk so that now most executives take it as read that they will type a good deal, perhaps all, of their written output. It took even less time from the introduction of email to the point where you are not seen as a serious player without it. New things may well consume us all even more quickly. We will see.

At the same time the pace of change does make problems. The cost of re-equipping or updating machinery, the training – formal or informal, it all takes time – and so on seems to continue in an endless cycle; and that is just one example. A new development may be real and useful, but to many people it sometimes seems to last about five minutes. The little ditty below summarizes the feelings of many people as they are faced with the next new gizmo.

*I bought a new computer;*
*It came completely loaded.*
*It was guaranteed for 90 days*
*But in 30 was outmoded.*

However, maybe such sentiments are possible only because of the computer industry's marketing success.

These examples may seem to be primarily in the area of office administration, but the electronic revolution has wide impact. Three areas deserve particular mention:

### 11.8.1.1 Products

The evidence for electronically influenced products is all around us. This includes things that are obviously electronic such as computers, computer games, digital cameras and personal organizers. It also includes things that appear just electrical, such as washing machines and fax machines, and others such as cars – how many chips are there in all of these, and more? More products in the future will be in these general categories. Maybe some of them will be yours.

The whole process of product development in these circumstances becomes very different to the development of simpler things. Original development may:

- take longer;
- cost more;
- be more complex (and thus more likely to be problematical);
- more vulnerable to competition.

We can all think of examples of the last of these. Facsimile machines saw off telex in a moment. Email has replaced most fax messages (less special fax paper sold, more normal paper as messages are printed out). Cassettes were largely sent on their way by compact discs, and a variety of music and video formats currently vie with each other to be the next 'standard'. This kind of dance is typical of many fields, from software to toys; only the timescale varies. Updating may follow very quickly. In some product areas new versions follow one another almost on a monthly basis, as with computers; on certain scales engineering is involved, for instance as more and more machinery is computer-controlled.

## 11.8.1.2   Methodology

The way things work is being changed by technology. Right across the business world, the range of change is enormous: think of the effect that electronic money transfer and hole-in-the-wall cash machines have had on bank branches; think too of the developing impact of Internet banking and where that may take us. Such things constitute big changes, and such an example barely scratches the surface of what is going on.

Things change everywhere. Customers used to go into a shop, select goods, pay, and that was the end of it. Paying now involves electronic machines at the cash points. These do not just facilitate the taking of money (and make it possible to employ staff who cannot add up!), but they are the tip of an electronic iceberg of integrated computer systems. The cash point registers the sale, communicates with stock control and more supplies of a product can be ordered automatically when stocks decline to a certain level. If a customer uses a card (increasingly so-called *smart cards*), whether a credit card or one linked to a loyalty scheme, then the sale can be recorded against an individual. Buy product X and the customer suddenly starts receiving promotions through the post for product X or its competitors (something like a supermarket can charge its suppliers to send this sort of material on their behalf). Such schemes can be linked back to the cash point, so that offer coupons are distributed to particular customers in a way that reflects their buying record – or rather, of course, in a way intended to influence their future buying.

Again, space prohibits a lengthy list of examples but the current complexity and future possibilities are clear. For any particular organization there are matters here that must be coped with. For example, a supermarket might be more resistant to seeing salespeople if its computer system is able to reorder directly. Such salespeople then have to find new ways of prompting the discussions they want, discussions that go a long way beyond reordering and involve promotion, display and much else, which are vital to the marketing effort.

As well as difficulties, there are also opportunities, areas where you can choose to get involved, despite greater complexity, if you see an advantage. For example, a search of what is new and an assessment of how it might help have become a prerequisite part of marketing thinking. Examples may well date, but the following show what is now possible, indeed what is now normal, in terms of going about things.

- Computers can now calculate optimum merchandising arrangements. A company making a range of products different in size, price, margin and rate of turnover can work out what mix of product selection should be put on any particular amount of shelving that a store allocates to its brand, at the touch of a button. No spare space is then left on the shelf and turnover and profit generation are maximized.
- Field sales staff now routinely carry computers and can use them for instance to give an instant answer to customer questions about stock and delivery of a product.
- Customer details can be accessed instantly during transactions just by the mention of, say, a postcode; this facilitates many processes, for example dealing at a distance from a call centre.

Computer and other IT developments will doubtless produce many more developments of this sort, all either actively assisting marketing effort, or with which marketing must fit in if it is to retain credibility.

### 11.8.1.3   Electronic marketing

The other major area, and one subject to continuing change and development, is the whole area of the Internet and e-commerce. These tend to be spoken of as if they were something new. In fact, though *how* they operate is clearly new, the effect they have is to add to the choice of methodology available in two different areas: promotion and distribution. Let us take these in turn.

- **Promotion:** Websites, for all their technical wizardry, are only another method of communicating with customers (and sometimes – see 'Distribution' below – of doing business). Consider a website, first and foremost, as a promotional channel. As such it must command attention, put over its message clearly and act to persuade. It must also be convenient and easy to use. At this stage, and for a while yet, this means appealing to people who see themselves to be other than at the forefront of computer literacy (see Section 11.8.2 'Help, please'). The website may be an electronic alternative to many things: a brochure, a salesperson, a showroom or shop window, a magazine or more (in any combination).
- **Distribution:** Here we are in the territory of what has become known as *e-commerce*. This is a situation where the whole business transaction, or most of it, takes place over the airwaves, as one might say. The point about clarity and convenience made above is perhaps even more important here. Some things are up and running and working well. Customers may conduct their banking and finances through Internet accounts; they may order many things – computers, pizzas, books, CDs and so on – from Internet sites. One point about this that will be worth watching is the relationship between Internet shopping and conventional retailing.

Some things can work fine exclusively through Internet channels. If someone wants to buy a new novel by a favourite author, they are probably happy not even to look at it first – they tap into amazon.co.uk or whoever, call up the title and place their order (with maybe a little price comparison along the way). Other purchases are more complex. If someone wants a new CD player, say, they may well want to look at it – better still, hear it and check it out. They go to a retailer and do just that. Then they might elect to visit a number of sites on the Internet, compare prices, check delivery and so on and place an order. What many people will not do is simply order a machine seen only on a monitor screen.

In other words, retailing is currently necessary for certain kinds of e-commerce to work. How this works out and what future buying practice will be like are still uncertain. That said, e-commerce works well for many things, and the fields in which it operates are growing and more and more people are either experimenting with this sort of shopping or expressing confidence in it and becoming regulars. Current predictions for the further future include the demise of supermarkets. All bulk goods, from tissues to cat food, will be ordered over the Internet and delivered. Customers will visit stores only for things that demand real choice or checking; as a result stores will be smaller, but departments such as the bakery or cheese counter will expand. Again, we will see.

### 11.8.2   Help, please

An important fact about the nature of websites is worth emphasizing: *they must be customer-focused*. This may seem obvious. But it also means that they should *not* necessarily:

- incorporate everything that is technologically possible (perhaps just because someone regards doing so as a challenge);
- be comprehensive (some things that might be incorporated are surely more important than others);
- be interactive in every possible way (though there may be strong reasons for having an interactive element);
- incorporate every technological gizmo known to man (some sites will aim for customers who are themselves more technically sophisticated or demanding; others must recognize that not everyone is of this persuasion; at least not yet).

It does mean that the objectives that give a website its *raison d'être* should be customer-oriented – it should be designed to work in the way they want, or at least find they like; and to do so without any great gaps in its capability.

An example will illustrate the point. One evening I contacted two websites. One was that of Amazon, the American bookseller (they, of course, sell more than books but it was a book order that prompted the contact). The site is especially clear and easy to use. All was going well. A few interesting minutes were spent checking out soon-to-be-published books in favourite areas and an order was placed. Then a problem materialized (about credit cards – the details do not matter). The system did not cope with this and the user's understanding did not cope with how to deal with it either. In a shop, of course, you would just ask. Same here – a message was sent, by email, and a prompt reply spelled out clearly exactly what needed to be done. No problem. When a site works this well people will return (even those who do not see themselves as at the forefront of things electronic).

The same evening another site was contacted. It was a nightmare of insufficient information and confusion (it would be unkind to name it), and, after some minutes of struggle, frustration and travelling in electronic circles, any attempt to do business with it was abandoned.

The difference was very obvious when the two sites were seen alongside each other, as it were. Why should one be so good, another so poor? After all, the technology is there; some people make it work and get customers saying, 'This is good!', whatever the technical sophistication of the site. A simple one is certainly able to do a good job and make customers feel it is good. It is probably in the approach. Maybe the second site had been set up too quickly. Maybe it had been created on the assumption that customers are clairvoyant or maybe the objectives for the site were unclear.

Whatever the reason, the point here is clear. Do business in this way and you expose yourself. No order materialized in this example. Assuming they contact you at all, people will notice how your site works (not least compared with others), they will talk about it to others and they will elect to come back – or not. Image is affected; so are future business prospects. Going about the setup process needs care and consideration, and it is to this that we turn next.

## 11.8.3   Going electronic

A complete rundown on the methods of marketing incorporating the Internet is beyond our brief here. However, some guidelines about the setup and role of a website follow, both as practical advice in its own right and as an example of the thinking and approaches that need to be applied in this area.

The job of setting up a website can be time-consuming and expensive; so too can be maintaining it and keeping it up to date. Some organizations acted very early as technology created this opportunity, although some acted solely because it was flavour of the month, something that 'had to be done', perhaps doing so to keep up with others, perhaps to pander to the ego of someone involved and enthusiastic. Whatever the reasons, there are certainly examples where such early action was ill considered (or indeed not considered), and where time and money were spent to no good effect. Such an initiative needs thinking through; the first question is very obvious and straightforward: 'What are the objectives that you have for your website?'.

It is not suggested that there will be only one. Several are likely, but they should all be spelled out and be specific. Ultimately, it will be important to know whether the cost of setup is delivering what was intended; and this is important to how a site is developed. Two particular purposes predominate.

1.  **A reference point:** Perhaps the site is in part a source of reference. You want people to consult it to obtain information (and be impressed by it at the same time). For example, many accountants have sites on which you can check current tax information or rates. Such may save time and effort otherwise expended in other ways. Perhaps you intend that a site will play a more integral part in the overall sales and marketing process. In that case, you will want to measure its effectiveness in terms of counting the number of new contacts it produces and, in turn, how many of those are, in due course, turned into actual paying customers.

    *Note:* If you already have a website check whether the organization gets good feedback on its use and the specific results it brings you (for example, counting new contacts or revenue coming from new contacts). Similarly if you are in the process of setting up a site – ensure consideration of this is an inherent part of the process.

2.  **An ordering point:** In addition, you may have products you want people to order and pay for through direct contact with the site. For example, a consultant might offer a survey of some sort, primarily to put an example of their expertise and style in the hands of prospective clients (though it might also be a source of revenue). A product company might, of course, have their whole range listed and available to order off the site. In this case, not only must the ordering system

work well, and this means it must be quick and easier for whoever is doing the ordering, but the follow-up must be good too. Any initial good impression given will quickly evaporate if whatever is ordered takes forever to arrive or needs several chasers. One hazard to good service is to demand too much information as an order is placed. Some facts are key and, of course, this kind of contact represents an opportunity to create a useful database; but turning ordering into an experience reminiscent of the Spanish Inquisition will hardly endear you to people.

### 11.8.4    Three distinct tasks

With clear objectives set, there are then three distinct tasks that must be fulfilled.

1. **Attract people to the site:** Just the existence of the site setup does not mean people will log onto it in droves, much less that the specific type of people you want to do so will act in this way. Other aspects of wider promotion must draw attention to it and this may vary from simply having the website address on your letterhead to incorporating mention (and perhaps demonstration) of it into a range of promotional methods, from advertisements to brochures. Simple, cost-effective methodology may work well here with just a simple promotional postcard acting to prompt people (customers or others) to investigate the site.

2. **Impress people when they see it:** Both with its content and its presentation. This means keeping a close eye on customers' views and accommodating all the necessary practicalities as it is set up. For example, all sorts of impressive graphics are possible. They can look creative and may well act to inform and impress. Certainly you will need some. But such devices take a long time to download and, if that is what you are encouraging people to do, they may find this tedious, at worst curtailing their contact because of its time-consuming nature. This is more likely if the graphics seem more like window dressing than something that enhances the content in a way that is genuinely useful to customers.

3. **Encourage repeat use:** This may or may not be one of the objectives. If it is, then efforts have to be made to encourage recontacting (again using a whole range of prompts) and this too may involve an overlap with other forms of communication.

In addition to the points made above, you will also need to consider carefully:

- **site content:** what should be presented (this is an ongoing job, not a one-off);
- **interaction:** how the contacting of the website can prompt a dialogue;
- **topicality:** how up to date it should be (this affects how often it needs revision, from daily to annually);
- **ease of use:** its convenience and accessibility (does it have a suitable navigation mechanism?);
- **image:** whether it will look consistent (and not as if it has been put together by committee);
- **security:** the protection it needs (is anything confidential, is it vulnerable to hackers, etc.? and will customers feel their own information is safe?).

Overall, it will need the same planning, coordination and careful execution as any other form of marketing communication. In addition, it is likely to necessitate active, ongoing cooperation from numbers of people around the organization who will provide and update information. This may be a larger job than it appears at first sight, not just because of the numbers of individuals and departments involved, but because they may have differing perspectives (with, say, research and marketing differing in the depth of technical information that should be included).

This aspect can present quite a challenge, especially in an organization of any size. Clearly, responsibility for the site and what it contains must unequivocally be laid at someone's door, together with the appropriate authority to see it through. Another challenge may be to get computer and marketing people to work effectively together. Marketing people must, for instance, ensure that computer staff understand the objectives and do not proceed on the basis of including everything that is technically possible (or, at the risk of upsetting computer experts, that is just fun to do).

At the same time, someone needs to have the knowledge that is necessary from a technical standpoint. This may be internal or external, but it needs to be linked to an understanding of marketing and/or the ability to accept a clear brief if an appropriate scheme is to be created. There is a real danger of simply applying all the available technology, building in every bell and whistle simply because it is possible. Practical solutions are necessary to meet clear objectives (and these should always be customer-focused).

If a site is to be useful, i.e. it is to comprise an effective part of the marketing mix, then sufficient time and effort must be put in to get it right. And the ongoing job of maintaining it must be borne in mind from the beginning.

## 11.8.5   Additional possibilities

### 11.8.5.1   Linking in research

An interesting and practical development is the recent availability of standard, cost-effective software packages that can work as an integral part of a website and monitor how it is used. In fact, there are now such add-ons better described as research tools. These typically allow regular research and formal monthly analysis about exactly who is using a website, their precise characteristics, and how and why they are in touch with the site. This allows the way the system works to be simply tailored to the needs and intentions of an individual user. The intention is specifically to obtain information that will make the website a more accurate and effective marketing tool.

### 11.8.5.2   Linking to sales

Similarly, there are now systems that allow a visitor to a website to click in a way that institutes their receiving a telephone call to discuss some specific detail of an offer. This can be instant and online, so that both parties can look at the site on screen and discuss it. Alternatively, a call can be made a few minutes later. It depends on the system. Such a contact is, of course, essentially a sales one. If it is made sufficiently easy, then it will generate conversations that can influence the likelihood of sales but that might otherwise never occur.

### 11.8.5.3   Utilizing appropriate technology

Some applications are particularly well suited to a specific product or service. For example, it is possible to book a hotel following a detailed inspection of it online. Before long customers will do this in a way that is almost as real as walking around the actual building. This parallels what many customers want to do. If a real alternative to actually visiting a hotel is provided, something that is judged by its users as better than any sort of brochure, then a particular provider will have an edge in the market against their competitors.

It is said that the future is not what it used to be. Certainly marketing people have a whole new area of activity, and new skills as well, to get to grips with. Precisely how it can – and will – affect you over the coming months and years may be uncertain. That it *will* affect you is not.

One sure way of being better at marketing in future is to get to grips with what needs to be done in this area. It needs careful consideration (it is not something to jump at) and whatever is done will need care in implementation. It is not something that can ever be got 'right', and put on one side as needing no more care and attention. For better or for worse, we are all faced with an ongoing process as change continues and new elements of all this come into view. It presents both a challenge and an opportunity, and marketing is inherently about the creative exploitation of opportunities.

## 11.9   Deciding the mix

It is probably clear from all that has now been said about promotion that selecting the 'right' mix and implementing the actual promotional activity are a complex task. So too is deciding how much to spend. Clearly, success, in terms of promotion of any sort that works, does not 'just happen': a systematic approach is necessary, and so is a degree of formality. The next subsection, 'Planning the promotional strategy', sets out a classic 12-stage approach, and Section 11.10, 'Setting the promotional budget', comments on that important issue.

### 11.9.1   Planning the promotional strategy

Any company must:

1. **analyse** the market and clearly identify the exact need;
2. ensure the need is **real** and not imaginary, and that support is necessary;
3. establish that the **tactics** they intend to adopt are likely to be the most cost-effective;
4. define clear and precise **objectives**;
5. Analyse the tactics available, taking into consideration the **key factors** in:
   - the market
   - specific target audiences

- the product range offered
- the company organization/resources;

6. select the **mix** of tactics to be used;
7. check the **budget** to ensure funds are available;
8. prepare a written **operation plan**;
9. **discuss and agree** the operation plan with all concerned and obtain management decision to proceed;
10. communicate the **details** of the campaign to whoever is implementing it and ensure that they fully understand what they must do, and when;
11. implement the **campaign**, ensuring continuous feedback of necessary information for monitoring performance;
12. analyse the **results**, showing exactly what has happened, what factors affected the result (if any) and how much the campaign cost; this stage may usefully include communication to all concerned, to complete the circle.

Nothing in the promotional plan must become automatic in the sense that a successful promotion is simply repeated without thought. Nothing lasts for ever.

## 11.10    Setting the promotional budget

There are several approaches to the complex issue of setting the promotional budget.

### 11.10.1    Percentage of sales

To take a fixed percentage – based, usually, on forecast sales – relies on the questionable assumption that there is always a direct relationship between promotional expenditure and sales. It assumes, for example, that if increased sales of 10 per cent are forecast, a 10 per cent increase in promotional effort will also be required. This may or may not be realistic and depends on many external factors. The most traditional and easiest approach, it is also probably the least effective.

### 11.10.2    Competitive parity approach

This involves spending

- the same amount on promotion as competitive firms; or
- a proportional expenditure of average industry appropriation; or
- an identical percentage of gross sales revenue compared with other similar organizations.

The assumption is that in this way market share will be maintained. However, the competition may be aiming at a slightly different sector and including competition in the broadest sense is no help. If you can form a view of competitive/industry activity it may be useful, but the danger of this approach is that competitors' spending represents the 'collective wisdom' of the industry, and the blind may end up leading the blind!

It is important to remember that competitive expenditure cannot be more than an indication of the budget that should be established. In terms of strategy, it is entirely possible that expenditure should be considerably greater than a competitor's – to drive them out – or, perhaps for other reasons, a lot less.

Remember that no two firms pursue identical objectives from an identical baseline of resources, market standing and so on, and that it is fallacious to assume that all competitors will spend equal or proportional amounts of money with exactly the same level of efficiency.

### 11.10.3   Combining percentage of sales and competitive advertising expenditure

This is a slightly more comprehensive approach to setting the budget, but still does not overcome the problems inherent in each individual method. It does, however, recognize the need for maintaining profitability and takes into account the likely impact of competitive expenditure.

### 11.10.4   What can we afford?

This method appears to be based on the premise that if spending something is right, but the optimum amount cannot objectively be decided upon, whatever money is available will do. Look at:

- what is available after all the other costs have been accounted for (premises, staff, selling expenses and so on);
- the cash situation in the business as a whole and the revenue forecast.

In some companies, advertising and promotion are left to share out the tail end of the budget – more expenditure being considered to be analogous with lower profits. In others, more expenditure on promotion could lead to more sales at marginal cost, which in turn would lead to higher overall profits.

Again, this is not the best method, demonstrating an *ad hoc* approach that leaves out assessment of opportunities in both the long and short terms.

### 11.10.5   Fixed sum per sales unit

This method is similar to the percentage-of-sales approach, except that a specific amount per unit (for example, per individual product sold) is used rather than a percentage of pound sales value. In this way, money for promotional purposes is not affected by changes in price. This takes an enlightened view that promotional expenditure is an investment, not merely a cost.

### 11.10.6   What has been learned from previous years?

The best predictor for next year's budget is this year's, asking:

- Are results as predicted?
- What has been the relationship of spending to competition?

- What is happening in the market?
- What effect is it having?
- What effect is it likely to have in the future?

Experiment in a controlled area to see whether the firm is underspending or overspending. Monitor results by tracking the awareness of promotions among customers, and the results of experiments with different budget levels can then be used in planning the next step (although you must always bear in mind that all other things do not remain equal).

## 11.10.7  Task method approach

Recognizing the weaknesses in other approaches, a more comprehensive four-step procedure is possible. Emphasis here is on the tasks involved in the process of constructing a promotional strategy, as already described. The four steps of this method are the following.

1. **Analysis:** Make an analysis of the marketing situation to uncover the factual basis for a promotional approach. Marketing opportunities and specific marketing targets for strategic development should also be identified.
2. **Objectives:** From the analysis, set clear short- and long-term promotional objectives for continuity and build-up of promotional impact and effect.
3. **Tasks:** Determine the promotional activities required to achieve the marketing and promotional objectives.
4. **Costing:** What is the likely cost of each element in the communications mix and the cost-effectiveness of each element?

Which media are likely to be chosen and what is the target level? (This is the number of advertisements, leaflets and so on.) For example, in advertising, the media schedule can easily be converted into an advertising budget by adding space or time costs to the cost of preparing advertising material. The promotional budget is usually determined by costing out the expenses of preparing and distributing promotional material or whatever. Here a variety of options may need considering, balancing greater or lesser expenditure against larger or smaller returns.

The great advantage of this budgetary approach compared with others is that it is comprehensive, systematic and likely to be more realistic. However, other methods can still be used to provide rough or 'ballpark' estimates, although such methods can produce disparate answers, such as:

- we can afford £10 000;
- the task requires £15 000;
- to match competition requires £17 500;
- last year's spending was £8 500.

Note: this just takes some nominal round figure; in some industries the amounts would be in millions.

The decision then becomes a matter of judgement, making allowances for your overall philosophy and objectives. There is no widely accurate mathematical or automatic method of determining the promotional budget. The task method does, however, provide, if not the easiest, then probably the most accurate method of determining the promotional budget.

In a large company, or one with a substantial promotional budget, this will be carried out by, or with, an advertising agency, with certain of the tasks such as media buying, the creative input, being carried out exclusively by them. It is not necessary, nor is there the space here, to explore this planning process in detail, but it must be done, and done well.

Even one simple (simple?) error can cause major problems. For example, you do occasionally see advertising for a new product, yet are unable to find it in the shops because the manufacturer has got the timing of advertising and stocking out of line. It can also happen with PR, as with a book review that appears ahead of publication. This kind of mistake cannot only be costly, but can also alienate customer loyalty for the future.

## 11.11   Summary

Promotion is, as we have seen, something that encompasses a number of techniques. These need to be deployed:

- in the right mix;
- against sound, well-coordinated objectives;
- in a *creative* manner.

Though fashion and copying are evident in promotion, and particularly in advertising, originality and creativity are two very important aspects for any successful promotion. A very creative and original scheme, even though inexpensive, can (and often does) score over a high-budget, stereotyped, uncreative approach. While advertising can never sell a poor product (certainly not more than once), a well-thought-out and consistent approach can become memorable.

Advertising slogans even pass into the language; at their best they can create an awareness of something, indeed a desire for it, which may be very difficult for a competitor to overthrow. So, perhaps even if half the money spent on promotion is wasted, the other half remains very important.

In any case, promotion – however good – can be wasted unless backed by effective sales and service, which are the topics of the next chapter.

*Chapter 12*

# Additional persuasive influences

In many industries (including many kinds of engineering), promotional activity rarely produces actual business. That is not to say that it is useless; its job is to create interest. Turning that interest into an actual order is the job of sales; the work of salespeople is the only part of the marketing process that involves direct, individual, personal contact. So selling, and how it and the customer base are managed, deserve some attention here, bearing in mind too that the people doing the selling may vary: in an engineering consultancy, for instance, it may include members of the professional or management team. Also deserving a mention here is another personal element directly affecting marketing success: the style and effectiveness of customer service.

## 12.1  Personal persuasion

### 12.1.1  The nature of the process

Now selling can sometimes have an unfortunate image. Think of your own instant judgement on, say, a double-glazing or insurance salesman. The first words that maybe come to mind are 'pushy', 'high-pressure' or 'con man'. Selling can be associated with pushing inappropriate goods on reluctant customers. Selling refrigerators to Eskimos is perhaps the kind of situation that springs to mind in those not directly involved in sales (though Eskimos *do* buy refrigerators – they need them to keep food *warm enough* to cook without defrosting! – but I digress).

The best – that is, most effective – selling is well described as 'helping people to buy'. Much of it has advisory overtones and, if it is to be acceptable as well as effective, it cannot be pushy, but must, like everything in marketing, be customer-oriented. Selling is, in fact, a skilled job and demands a professional approach of the field-force personnel who carry it out. Customers may want the product, but with plenty of alternative sources of supply they are demanding, and persuading them to do business with a particular supplier may be no easy task.

In most businesses, selling must take place if the marketing process is to be successfully concluded. At one end of the scale it is simple: just a question. As an example of selling in its simplest form, a store selling spirits may be able to increase sales significantly just by ensuring that, every time a member of staff is asked for spirits, they ask, 'How many mixers do you want?' Many people will respond positively to what has been called the *gin-and-tonic effect*, the linking of one product with another. Sometimes the question is even simpler: the waiter in a hotel or bar, for example, who asks, 'Another drink?' is selling. Such should be easy to implement – just an instruction to staff. And it is the same with spare parts for many engineering items.

At the other end of the scale, sales do not come from the single isolated success of one interaction with the customer. A chain of events may be involved, several people, a long period of time and, importantly, a cumulative effect. In other words, each stage, perhaps involving some combination of meetings, proposals, presentations and more meetings, must go well or you do not move on to the next.

So, with the thought in mind that the detail of what is necessary will vary depending on circumstances, let us review both the stages in turn, and some of the principles involved throughout the sales process.

Selling starts, logically enough, with identifying the right people to whom to sell. Sales time is expensive, so it is important for salespeople to spend time with genuine prospects, the more so when the longer lead time referred to above – and typical in the purchase of, say, computer systems – is involved. A salesperson in some industries can spend most of their time with regular contacts – as is the case with publishing, where bookshops receive regular calls. Otherwise, a constant supply of new contacts may be necessary. Some of the right people come forward as a result of promotional activity. They phone up, return a card from a mailshot, or whatever and, in so doing, are saying, 'Tell me more'. Others have to be found; finding them is the first stage of the selling process.

## 12.1.2   Prospects – the process of location

An ongoing process of prospecting must, where necessary, produce a constant stream of new names. A variety of sources can be tapped: everything from directories and trade bodies to exhibitions and the press. One or a combination of these can supply valuable information about prospects: the names of companies, what business they are in, whether it is going well or badly, whether they export, how big they are, who owns them, what subsidiaries or associates they have and, last, but certainly not least, who runs and manages them.

Exactly which individual is then approached is obviously vital, and may not be a simple decision. Indeed, it may be that more than one person is involved. For example, for business clients of a travel agent, the following must be considered: the person who travels, the person who sends them, the person who pays and perhaps also the person who makes the booking. There is many a secretary with considerable discretionary power in making company travel bookings, and not least among their considerations will be how straightforward and easy the travel agency is for them to deal with personally.

As well as considering which individual to approach, the other important assessment at this stage is that of financial potential. How much business might be obtained from a prospect in, say, a year? This analysis will rule out some prospects as not being worth further pursuit. Experience will sharpen the accuracy with which these decisions can be made, but, meanwhile, a good first list is developing.

The old military maxim that 'time spent in reconnaissance is seldom wasted' is a good one. In war it can help to prevent casualties. In business it not only produces information, in this case on who should be contacted, but it also provides a platform for a more accurately conceived, and more successful, approach.

So, having identified who will be contacted, the next step is organizing the approach. A number of factors may be important here, both before an approach is made and in follow-up. Two key areas that need to be considered before making an approach are selecting the method of approach and deciding who will take the action. Let us look at both in turn.

### 12.1.2.1   Selecting the method of approach

The ultimate objective is almost certainly a face-to-face meeting, which must be held before any substantial business can result. Such a meeting can be set up by such methods as:

- cold calling, that is, calling without an appointment;
- sending a letter, card or email (not spam) with or without supporting literature;
- telephoning cold or as follow-up to a letter or promotion;
- getting people together, initially as a group, and making a presentation at your premises, a hotel or other venue, or through a third party (such as at a trade-body meeting) and more.

The logistics are also important. What is needed is a campaign spread over time so that if and when favourable responses occur they can be followed up promptly; such responses may be more difficult to cope with if many occur together.

### 12.1.2.2   Deciding who will take the action

The prospecting process will almost certainly involve approaching, meeting and discussing matters with people senior in, and knowledgeable about, their own business. The approach therefore needs to be made by people with the right profile, who will be perceived as being appropriate, and who can really give an impression of competence. They will also need to have the right attitude, wanting to win business in what may be a new and perhaps more difficult area. And they need the knowledge and skills to tackle the task in hand: knowledge of the customers, their own organization and its products and so on, backed by more technical elements where appropriate.

Detail is important. Returning to the example of a travel agent, the export manager who is made late for an appointment will be equally upset whether he has missed a flight connection because he overslept or has been misinformed on the time it takes to get from airport to hotel. The travel agent, rightly or wrongly, will probably get the blame. Finally, skills in customer contact, selling and negotiation are needed as well as in all those areas – such as writing sales letters – involved in making the approach.

The initial approach is vital, like any first impression, and it may be very difficult, having received an initial negative response, to organize a second chance. If you have thought the process through in this way, the chances of success are that much greater.

However it is set up, once contact occurs, the salesperson has to make and carry through a personal contact and to do this must understand the potential buyers and make contact with them persuasive yet acceptable; in other words, not so 'pushy' as to be self-defeating.

## 12.1.3   The sales process

Selling goes through various stages. It starts, as has been said, with an understanding of the buyer. No one can sell effectively without understanding how people make decisions to purchase. A good way of thinking about it, one originated by psychologists in America, suggests that the decision making that goes on in the buyer's mind goes through seven distinct stages, which may be paraphrased thus:

1.   I am important and I want to be respected.
2.   Consider my needs.
3.   How will your ideas help me?
4.   What are the facts?
5.   What are the snags?
6.   What shall I do?
7.   I approve.

The customer must do what they feel is necessary to make a decision; indeed the way selling is conducted must allow this to happen.

The two keys to success – 'closing the sale' as obtaining a buying commitment is called – are, first, the process of matching the buyer's progression through the decision-making process, and, second, describing, selectively, the product and discussing it in a way that relates to precisely what a (particular) buyer needs.

There are a variety of techniques and skills involved here, and it is beyond our remit to review them all here, though the following gives a flavour of what is involved. If you do have what amounts to sales responsibilities, then perhaps I can mention (surely allowed in this chapter) that I have written separately on the subject in *Outsmarting your Competitors* (Marshall Cavendish).

Overall, a sound sales approach – or at least the key aspects of it – can be summarized. First, the basics. To be successful, field sales staff must be able to achieve the following.

- **Plan:** They must see the right people, the right number of people, regularly if necessary.
- **Prepare:** Sales contact needs thinking through, the so-called 'born salesperson' is very rare. The best of the rest do – and benefit from – their homework.

- **Understand the customer:** Use empathy, the ability to put themselves in the 'customer's shoes', to base what they do on real needs, to talk benefits.
- **Project the appropriate manner:** Not every salesperson is welcome, not everyone can position themselves as an adviser or whatever makes their approach acceptable. Being accepted needs working at.
- **Conduct a good meeting:** Stay in control, direct the contact, yet make customers think they are getting what they want.
- **Listen:** This is a much-undervalued skill in selling.
- **Handle objections:** The pros and cons need debating. Selling is not about winning arguments or scoring points.
- **Be persistent:** Ask for a commitment, and, if necessary, ask again.

Are there special skills that the sales staff must deploy? Yes, in most specialist companies, there certainly are. In publishing, many factors are important, for example with an extensive product range (titles may number thousands) acquiring the necessary product knowledge is more difficult than for those in a company selling a single product, especially if it is not technical. The most important factor is probably that of time. The time buyers allow publishers' representatives (as they are called) is small, so the amount of time that can be spent describing a single title may be very small (one to two minutes is not untypical).

Such factors are often crucial – here it is true to say that if someone cannot deal with things succinctly yet make a powerful case for them, they literally cannot do the job. Similarly, someone selling drilling equipment told me most of his meetings were held a mile underground with a lamp on his helmet! Again, think of what your own organization does.

Second, a variety of additional skills may be necessary to operate professionally in a sales role. These include:

- account analysis and planning;
- the writing skills necessary for proposal/quotation documents to be as persuasive as face-to-face contact;
- skills of formal presentation;
- numeracy and negotiation skills.

And all this in a job in which some people say of someone that they are 'only in sales'. Furthermore, whether a customer buys again and buys more is dependent primarily on two additional factors: service and follow-up.

- **Service:** It almost goes without saying, but promises of service made by salespeople must be fulfilled to the letter. If they are not, the customer will notice. A number of different people may be involved in servicing the account. They all have to appreciate the importance and get their bit right. If the customer was promised information by 3.30 p.m., a brochure in today's post, two suggested proposals in writing and a call to follow the whole thing up, then they should get just that. Even minor variations, such as information by 4 p.m., do matter. Promising what can be done and doing it 100 per cent are very much part of selling.

- **Follow-up:** Even if the service received is first class, the customer must continue to be sold to after any order, as contact is maintained in a variety of ways (without its seeming pushy).

A positive follow-up programme of this sort can certainly potentially make marketing work better. It can maximize the chances of repeat business and ensure that opportunities to sell additional products or services are not missed. Such an approach brings us full circle, from identifying and contacting prospects for the first time to holding and developing their business on a continuing basis.

But selling – certainly *effective* selling – does not just happen. The management process that is responsible for it is also important.

## 12.1.4  Sales management

It is not enough for a company simply to push salespeople out into the field and say, 'Sell'. As much as anyone else, salespeople need managing – why else have a sales manager, as most organizations of any size do? Usually such a manager is very much part of the overall management and marketing team. In a small company they may have many tasks and responsibilities that are really general management functions.

In addition, the person managing the sales team will usually handle a certain number of customers, usually larger ones, personally. There is nothing wrong with this, indeed such involvement is useful, but it can dilute the time available for the classic sales management functions and this, in turn, can leave sales less effective.

Effective sales management directly affects sales results and thus marketing success. Most usually the classic tasks of sales management are regarded as falling into six areas; these reflect the need to:

- **Plan:** Time needs to be spent planning the scope and extent of the sales operation, its budget and what it will aim to achieve. Achievement is organized first around targets, and setting targets not just for the amount to be sold but also for profitability, product mix, etc., is a key task. If the product range is large, this makes it especially important that the team's activities are directed with the right focus.
- **Organize:** How many salespeople are required needs calculating (not just a matter of what can be afforded, but of customer service and coverage; though the two go together), as does how and where they are deployed. This must also address the question of the various market sectors involved, looking at not just who calls on customers in, say, Hertfordshire, but how major accounts are dealt with and the strategy for any nontraditional outlets that may well need separate consideration. Organizations who sell to groups of customers that differ radically from each other may separate the different sales tasks, and even have separate sales teams.
- **Staff:** This is vital; it is no good, as they say, 'paying peanuts and employing monkeys'. If the sales resource is going to be effective then it must be recognized that recruitment and selection need a professional approach and the best possible team must be appointed.

- **Develop:** Here the process is ongoing. Because there is no one 'right' way to sell, what is necessary is to deploy the appropriate approach literally day by day, meeting by meeting, customer by customer and continue to fine-tune both the approaches and the skills that generate them over the long term. If the team is to be professional in this sense, then more than a brief induction is required: an ongoing continuum of field development is necessary.
- **Motivate:** A constant process that ensures that people not only can do what is required, but will want to excel.
- **Control:** As with any other kind of management, the need to control and fine-tune action is also important.

All in all, sales management has a wide and vital brief. The quality of sales management is often readily discernible from the state of the sales team. Excellence in sales management makes a real difference. Time spent on all aspects of sales management can have direct influence on sales results.

## 12.1.5   Sales productivity

One further area is worth mention. Whatever quality is brought to the conduct of sales calls, more than this influences sales results. The other crucial variables are:

- who is seen (the selection of appropriate prospects/buyers/customers);
- how many people are seen;
- how often they are seen (the call frequency decided upon and how it varies and is used).

If these productivity factors are well organized, and worked at on a regular basis, then the overall results are likely to be improved. It is a differentiating factor between the good and less good salespeople in any industry, and applies, albeit in slightly different ways, to all categories of customer, depending on the nature of the business.

Sales productivity comes first from seeing the right people, and the right outlets; and also the right individuals throughout larger outlets where salespeople may have to see several people in different departments (as in selling, say, training services to a company where purchases might be made by Personnel, Training or functional departments).

Second, it comes from sheer quantity. Provided call quality is not sacrificed, then, generally speaking, the more people are seen the more will be sold. And, third, there is the question of frequency. The rule here is often stated as to call the *minimum* number of times that will preserve and build the business. Some accounts are called on every week. Others may be seen only once a year. Whatever the circumstance, judging the right frequency of calls needs careful thought.

Overall, selling is a key constituent part of marketing, and one where regular and suitable thought can help make it work better; indeed, it is another area that demands constant fine-tuning in a dynamic marketplace.

Monitoring the numbers involved all along the line can influence action and results. For instance targets for the number of prospecting calls to be made, the number of sales calls to be scheduled and conducted, the frequency of calls on regular customers and so on are prescriptive: they prompt action that will make it more likely that final targets – the amount of product sold – are hit as a result.

## 12.2   Customer account management

Not all customers are the same. The way most marketers categorize them is by size: the amount of actual or potential business they produce. Major accounts then need special treatment. An extreme example is the grocery business. In the UK there are five major supermarket chains. One of those might be worth 25 per cent of the business of a company selling FMCG products; that is big however you look at it. Thus:

- Companies have people with titles like 'major account manager' who are responsible for the company's business with just a very few, but major, customers.
- Major-customer strategies are developed. In other words a plan for creating, maintaining and building a relationship with major customers is planned (and may be part of the overall marketing plan).
- A programme of regular contacts is scheduled for such customers and efforts made to make this fit the service and communication that the customer wants.
- Efforts are made to create a partnership between supplier and customer, with an emphasis on working together rather than simply 'selling' because the two businesses are inherently linked.

This happens in both industrial and consumer markets; indeed, in any enterprise the continuity and nature of communications and contact are crucial to developing business.

At the other end of the spectrum, some customers are small. If there are a lot of small ones, then together they may still have value, but techniques to deal with them recognize their nature and the cost of dealing with them. A personal visit may not be cost-effective; in these circumstances other methods can be used, for instance:

- **telephone contact**, particularly to ensure appropriate and timely restocking of a product (e.g. as a paper manufacturer might contact a small printer); call centres too are a sign of this strategy;
- **channelling** business through a wholesaler;
- leaving contact to the **customer** (e.g. as with a small shopkeeper visiting a cash-and-carry warehouse);
- allowing business only via a **specialist technique** such as mail order.

All this is relevant when a major customer is a regular one, as a supermarket or a chain of tyre fitters is. Other major customers may have a more one-off nature. For example, an engineering company may sell a water-treatment plant to a city, and then does not expect to sell them another very soon. This is more commonly called *big-ticket selling*, and is characterized by long lead times and high unit value.

### 12.2.1   Customer relationships

Whatever the precise nature of customers, an ongoing dialogue must be created with them. The contact must:

- suit the customer;
- be cost-effective;
- maintain active contact and focus on developing the business.

Marketers talk about 'relationship management' rather as if it were a generic term. In fact the form it can take may range widely. Amazon has excellent customer-relationship management techniques via its website, though it never meets its customers at all. Others do so via mail and catalogues. Still others have many personal contacts to arrange and see 'working with customers' as the way forward.

However it may be done, customers are demanding, fickle and sometimes difficult. The customer–supplier relationship must be managed in a way that they like (not simply how it is convenient to the supplier to act), one that attempts to screen out competition and build a strong business relationship. Despite best efforts, such relationships are fragile – and customers can and do change and begin to favour another supplier if they do not feel the whole thing is organized in a way that is best for them.

The quality of what is essentially a sales relationship is obviously affected by the service involved and delivered in the course of business being done. That must be right too, so it is to this we turn next.

## 12.3   Customer care

Looking after customers is inherent to marketing success. Customer service, as customers both anticipate and experience it, is fundamental to success. It is not overstating it to suggest that this is a basic foundation that underpins ongoing overall marketing success. The whole process involved here is nowadays most often called *customer care*, and its careful execution – in a way designed to impress customers – provides a significant opportunity to differentiate more effectively an organization from competitors.

### 12.3.1   Excellence in customer care

We have already seen that it is increasingly difficult for customers to differentiate between many of the competing products and services in the market. In many industries, products are essentially similar in terms of design, performance and specification, at least within a given price bracket. This is as true of industrial products as of consumer goods. Often customers' final choices will, therefore, be influenced as much by subjective areas as anything else. The nature of customer service can play a major role in this, sometimes becoming the most important factor.

The precise nature and manner of customer service adopted become an integral part of producing an organization's image. Indeed it should both reflect any existing image and extend it; so, if a company wants to be seen as efficient, modern, innovative or whatever, then its customer service must reflect such qualities. A company positioning itself as caring or advisory (in healthcare or financial services, for example) cannot skimp on time spent with customers without being thought unprofessional. Even little things can dilute an image. Consider something as simple as the effect of a direct mail shot, or simple sales letter. It may do a good job, sound right, prompt interest and commend urgency – but be let down by the fact that the reply-paid envelope enclosed with it is second-class (something that in the UK can routinely take four or more days to arrive). Even choosing such a simple example makes a point – the greater damage of worse mistakes is obvious.

Conversely, good customer-care practice can have a specific positive effect, binding a customer and supplier for long periods. The danger of the negative here – and everyone no doubt has their favourite horror story of bad customer service – is all too clear.

### 12.3.2   The opportunity of prevailing standards

Any shortfall in prevailing standards presents an opportunity for others in an industry to steal an edge by getting it right – one that surely is not so difficult to take advantage of and use to create a better edge in the market.

It is not a question of aiming for some idea of perfection (after all, both McDonald's restaurants and the Ritz Hotel would claim to offer good service, but in very different ways – and they are both right to say so). But any company must organize things positively to achieve the standards they have decided they – or rather their customers – require.

So what creates good customer care? It comes primarily through the careful consideration of both staffing and organization. It is not easy. The mix of characteristics and considerations that can help make success more likely is not easy to define; what we can be sure of is that customers know all too readily what they like – and the reverse – when they encounter it. Consider the positive and the negative.

### 12.3.3   Handling complaints

The first task is to reduce the number of complaints that must be dealt with. Every organization will doubtless get some, but everything that can give rise to them – the product or service, customer service and such factors as delivery, policy and people – must be examined to ensure that no complaints are occurring (especially with something happening repeatedly) that could be avoided.

Complaints are a source of feedback and intelligence. The job is not simply to field them and forget them. It is to learn from them and use what is learned to make things better for the future. Simple feedback and reporting systems may be necessary here, collecting and centralizing data so that they can be analysed and lead to positive action to improve things. Complaints must be handled in the right way, to create a positive out of a negative (again full how-to details are beyond our brief).

### 12.3.4 What customers want

To summarize succinctly here I will use a mnemonic much favoured by trainers. This says that customer contact – whether face to face, on the telephone, whatever – should be PERFECT. This is how the acronym works.

Polite: This almost goes without saying. It reflects the nature of the customer relationship with a supplier and must be manifest throughout the process of communication with them. It must be genuine: grovelling will reverse the desired effect; a pleasant personal touch will enhance it. And it must be maintained whatever the circumstances and the pressures.

Efficient: Things have got to be done right, and that means manifestly for the customer's convenience, not to fit in with an organization's systems, particularly bureaucratic ones.

Respectful: This is important and must match the customer. Some want this to be much in evidence, others less so, but it must always be there. (An understanding of their attitude to time is a good example: do they want everything done in a moment because they are in a hurry, or do they see time being spent on something as a show of thoroughness or care?)

Friendly: The level here must be judged just right – not every customer wants too much of this too soon, though everyone wants the transaction to be pleasant.

Enthusiastic: This displays an interest in the customer, which is something most regard as a prerequisite for good customer care.

Cheerful: Even in the face of adversity. Certainly no customer wants to feel that the whole business of their being dealt with is getting someone down.

Tactful: Many customer situations have aspects that are confidential or sensitive and respect for any such aspect is appreciated.

Overall, the above sums up the style of handling that works best; the final trick is to apply it individually. Customers like to be dealt with as individuals for the very good reason that that is what they are. Anything that smacks of a standard approach – dealing with them on 'automatic pilot' – dilutes the good that customer contact can do.

There are plenty of opportunities to make what is done here better than how competition operates, and gain a marketing edge from the process.

### 12.3.5 Managing to produce good customer care

Excellence of customer care does not just happen. Someone must manage the process and the people involved. Whoever does so has a job that involves a number of different elements. Thus …

- **It is not enough** for the manager responsible for support areas only to be a good administrator, although without the sorting out of priorities, without smooth handling of enquiries, files, paperwork, correspondence and records, sales support will never be effective.
- **It is not enough** for them only to be a good salesperson, although it is essential to have an understanding or familiarity with sales techniques, be able to recognize

sales opportunities, and ensure both they and members of their team meet them.

- **It is not enough** for them only to be an effective manager of people, although it is vital to be able to lead and motivate a close-knit and enthusiastic team, tackling a diverse range of activity in hectic conditions.

The manager also has to understand and pass on an understanding of the role of sales support, so that all concerned see it as a vital tactical weapon in the overall marketing operation. This means that they must have an appreciation of what marketing is and the various ways in which, directly or indirectly, the sales office and sales-support or customer-care staff can contribute to company profitability.

This implies a knowledge of, and involvement in, the marketing process. For example, if sales-support personnel are not told (or do not ask about) the relative profitability of different products, they may be busy pushing product A when product B, similar in price or even more expensive, makes more money.

In one company, for instance, sales office staff spent time handling complaints about delivery on 75 per cent of orders that went through. This was not because delivery was bad, but because members of the sales force – intent on impressing customers and getting orders – were quoting six weeks' delivery when everyone in the company knew it was normally eight weeks. Such unnecessarily wasted time could be used more constructively to increase sales. This example is all too typical.

### 12.3.6    Backup activities to support sales and service

Sales and service staff may succeed or fail largely through their own personal approaches, but they are necessarily dependent on the quality of the product or service that they sell, and the image of the company for which they sell it. So, if any job is involved with anything to do with either, then it helps influence sales success.

This is more than saying that those on the production line are involved: there are many more specific circumstances. A few examples of how people around an organization, who might just feel they have little or no relationship with 'marketing', can influence things are shown in Panel 12.1.

---

### Panel 12.1 Contributing to marketing

Examples of how people around the organization can contribute to marketing include:

- someone in technical support, handling a customer query, will not only sort out the problem, but influence the likelihood of a customer's reordering;
- someone responsible for originating a computer system that ultimately interfaces with customers will affect company image and thus the salesperson's relationship with their customers;
- someone in accounts, communicating to sort out some complexity of VAT on a customer account affects their image of the company for good or ill.

In some such areas the normal expectation, and experience, of the customer are that any good impression of the company will be diluted. Which customer, hearing the words 'it's in the computer' does not expect some inconvenience at their end? What are the equivalents of all this in your own organization? What is actively done to impress customers? What are the danger areas where problems might occur that customers would not like? Many people in different roles are involved in this sort of thing. Similarly, think about organizations with which you deal – what do you like and not like about their service and attitudes?

The marketing-oriented organization loses no opportunity to maximize the impression given by both customer care and sales. They overlap and both hold the possibility of assisting to make marketing better able to fulfil its intentions.

## 12.4   Summary

Part of the marketing process is very personal; overall marketing and promotional activity can achieve only so much. This means that:

- For many companies personal selling is a prime element in the marketing mix.
- Selling, which involves its own body of techniques, must be well conducted and the people doing it well qualified and well managed.
- Customers are not all the same. Particularly major customers need special attention, and tailored tactics must be used to develop their business.
- All customers, however, do expect good service and, although this is chrono-logically well down the chain of marketing events, it is also fundamental: good service can create customer loyalty and poor or bad service can guarantee that a customer votes with their feet and never orders again.

# Afterword

There you are. You should now (assuming you have reached this point by reading through from the beginning!) have a better understanding of the world of marketing, in concept and in its activities. We began with the three Ps: product, price and presentation, and added what is sometimes referred to as the fourth P, place, in Chapter 6, which dealt with distribution (and more besides). We have reviewed a wide range of topics, dissecting the process while at the same time stressing the nature of marketing as a cohesive whole.

So, given the nature of marketing, it is useful to end on a note linked to the overall concept again, rather than simply reaching the end of the dissection in the last chapter and stopping. To end, therefore, let us highlight five overall factors vital to marketing success, which we might call the five Cs (some were touched on in Chapter 6 with regard to distribution).

The first C of marketing is **customer**. The whole of every aspect of marketing, the concept, the planning through to all the research, communications and the application of every technique must focus on the customer. The customer is king, as the saying has it – and ultimately he who pays the piper calls the tune.

The second is **continuous.** Marketing is not an option, a bolt-on activity, or for moments when time allows. It must be present all the time as the company goes about its business. Indeed, without marketing there is a good chance it will *not* go about its business, at least not for long.

Next comes **coordinated**. Unless the many and various techniques of marketing are made to act together their effectiveness will be diluted. The different factors, sales and advertising – to pick two obvious ones – are not alternatives: it is likely that both are necessary. When, how and how much they interrelate and overlap is down to the skill of those involved.

Often this ongoing need for coordination is made more difficult by the number of people involved, usually spread across many departments and, sometimes, spread geographically – at its broadest, across the world. Marketing is, in the true sense, a management function.

The fourth C is **creativity**. Above all, marketing must differentiate, and, in what seem to be ever more competitive times, this is a challenging task. It is this combination of competitiveness and creativity that makes marketing so dynamic. It is not an exact science, and many of the variables are external. One never knows, for instance, what a competitor will do next. So, however the necessary creativity shows itself – through product innovation, clever (or, more important, persuasive and memorable) advertising or special attention to some aspect of service – it must always be present, bringing something new to bear to combat unpredictable competitive pressures.

And last, and by no means least, comes **culture**. Marketing is, above all, dependent on people – not only the people in marketing, the researchers, the marketing and brand managers, sales managers and account executives, but the many others throughout any organization who are, in fact, involved in some way with the process. Some are involved in an obvious way like those in customer care, since anyone who handles a customer enquiry (or complaint) or provides information, technical support or after-sales service contributes. Others, as we have seen, are a little more removed, but can still contribute. What is more, senior management in any organization should not only recognize this, but also work to ensure everyone contributes knowingly and positively; better still, that as they do so they understand why and get satisfaction from the contribution they are able to make.

The starting point of all this is an understanding of the company and its customers in a marketing context – and how it affects each individual. If your investigations of marketing, and your reading of this book in particular, help with that process of understanding, then you will have made a practical contribution to the process, because, ultimately, a marketing-oriented culture can help generate success and produce the financial strength (for many this is profit, which pays everyone's salary) that can only ever come from the market outside the organisation.

All five Cs are as important to engineering in all its forms as to any other discipline. The intention here has not been to set out a how-to-do-it guide for marketing in an engineering context. If this book has made one thing clear it is surely that marketing is a complex and creative process. It is one that must be applied specifically to fit any individual business, and certainly the job of marketing an engineering consultancy of some sort is very different from that of an industrial firm in heavy engineering.

But it is also a process that ranges widely, one that influences the overall success of any organization that stands or falls on its relationship with the market. And one too that, as you will have seen, is inherently linked to almost every other aspect of an organization. As such it affects many people, and many people can have influence on it and its degree of success.

If you are one such, then I hope you now understand better why this is so and what the opportunities of marketing are in your own organization.

# Index